THE LIFE OF A FOSSIL HUNTER

Charles H. Sternberg.

THE LIFE OF A FOSSIL HUNTER

CHARLES H. STERNBERG

Indiana University Press
BLOOMINGTON AND INDIANAPOLIS

FIRST MIDLAND BOOK EDITION 1990

Originally published in February 1909
by Henry Holt and Company

The paper used in this publication meets the minimum requirements of American National Standard for Information Sciences—Permanence of Paper for Printed Library Materials, ANSI Z39.48-1984.

∞™

Manufactured in the United States of America

Library of Congress Cataloging-in-Publication Data

Sternberg, Charles H. (Charles Hazelius), 1850–
The life of a fossil hunter / Charles H. Sternberg.
p. cm.
Reprint. Originally published: New York : Holt, 1909.
ISBN 0-253-35549-4. — ISBN 0-253-20571-9 (pbk. : alk. paper)
1. Sternberg, Charles H., (Charles Hazelius), 1850–
2. Paleontologists—United States—Biography. I. Title.
QE707.S8A3 1990
560.9—dc20
[B] 89-38604
CIP

1 2 3 4 5 94 93 92 91 90

CONTENTS

ILLUSTRATIONS

FOREWORD

By Rudolf A. Raff

On a warm day in June of 1876 the river steamer *Far West* began a hurried journey down the Bighorn River to the mouth of the Yellowstone, carrying the news of the destruction of Custer and much of his Seventh Cavalry by the Sioux at the Little Bighorn. The shocking news that came down river reverberates even yet in the mythology of the American West.

The American West in the nineteenth century held the bright vision of Manifest Destiny, the expansion of the United States to the Pacific. It represented the vast open frontier, free homesteads, adventure, wealth, growth. The steam-driven boldness of technological America was expressed in the westward thrust of the railroads and the probing of paddlewheel steamboats into the upper reaches of the Missouri. American intellectual development too was tied to the West. Francis Parkman, John Charles Frémont, and Washington Irving created a thirst for western adventure.

Artists found their abilities tested in developing vision and techniques to express the size, the strangeness, and the color of the new country. Vastness and light appear in the canvasses of Alfred Bierstadt and Thomas Moran. The Indians assume colorful but alien life in the portraits of Karl Bodmer and George Catlin. These images and others—savage mountain men, carefree cowboys, psychopathic gunfighters, lonely prospectors, hardy settlers—are all still part of the tradition of the Great West, and provided the movie fare for all those twentieth century generations that grew up before *Star Wars*.

There was another less romantic intellectual theme as well. The West was an invitation to scientific discovery on a scale that today can scarcely be imagined. Much of American science was born there. The U.S. Cavalry and the Sioux were not alone on those barren hills of southeastern Montana during the summer of 1876. In the few weeks following the Battle of the Little Bighorn, a party of three dinosaur hunters, led by the indomitable Quaker paleontologist Edward Drinker Cope, was also in the field. Cope reasoned that the Sioux would be kept distracted by the Army and consequently that war parties would simply not be a risk. Cope was right. The scientists met few Indians, and those who did visit the scientists' camp were treated to the amazing spectacle of Cope removing and replacing his false teeth. The recorder of these events was Charles H. Sternberg. His adventures with Cope mark the beginning of a sixty-year career as a pioneer collector and preparer of vertebrate fossils. Although he had no university degrees, Sternberg was an accomplished man, both as a scientific explorer and as a marvelous writer. His autobiography, *Life of a Fossil Hunter*, records the unique story of his life in the West.

Organized science in the West began before the Civil War with geographical explorations conducted by the elite Army Corps of Topographical Engineers. These explorations were mounted with the practical goals of mapping the West, defining the frontiers of the newly won empire, and finding routes for settlement. With the end of the Civil War, the demand for information about the West grew. The need for geographical information continued to be important as the railroads were built, and there was a growing demand for knowledge about resources. This was the age of mining, fabulous silver strikes, and outrageous mining barons. Proper surveys of the geology of the West were needed. And the public had a great eagerness to hear more about such rumored wonders as the geysers of Yellowstone and the ancient cliff dwellings concealed in the desert canyons of Navajo country. All of these interests would be fulfilled in elaborate government-sponsored surveys that took with them geologists, surveyors, artists, photographers, zoologists, and botanists.

Initially there were several survey organizations. Two surveys were sponsored by the Army. One was headed by Lieutenant George Montague Wheeler; the other was led by Clarence King, who had surveyed the mountains of California and wrote *Mountaineering in the Sierra Nevada*, a witty and lively account of his experiences with the California Survey. The two surveys sponsored by the Department of the Interior were headed by Ferdinand Vandeveer Hayden and by John Wesley Powell. In 1879 all of these activities were consolidated under the newly formed United States Geological Survey, which remains an outstanding research organization to this day. The leaders and many of the participants of the western surveys were amazing characters. Hayden, for instance, was so intent on collecting fossils that the Sioux regarded him as crazy and called him "the man who picks up stones running." Powell, who as the major of an Illinois artillery battery had lost an arm at the Battle of Shiloh, organized and led the first traverse by boat of the Grand Canyon of the Colorado in 1867. His account, *Exploration of the Colorado River*, remains a classic of exploration and adventure. He later headed the U.S. Geological Survey and the Bureau of American Ethnology.

It was with the wealth of materials returned by the surveys that the systematic study of western paleontology began. Some of the fossil finds from the West were to play important roles in providing evidence in favor of evolution. A whole series of fossils illustrating the evolution of the horses is a notable example. The discovery of the existence of dinosaur bones in the West soon followed and led to an unparalleled fossil rush.

In many ways the last quarter of the nineteenth century was an age of outsized figures. This was to be very much the case where dinosaurs were concerned. American vertebrate paleontology was dominated by two mutually antagonistic giants, Edward Drinker Cope and Othniel Charles Marsh, who launched the heroic era of American paleontology. Their discoveries and their feuding became legendary. Both men had wealthy relatives as benefactors and were therefore able to afford elaborate private expeditions. Both employed agents in the West to scout out fossil beds, and to keep an eye on the competition. There are stories of collectors offering their services under assumed names, of fist fights between rival collectors, and even of a stolen boxcar of fossils. Both men rushed descriptions of new dinosaurs and other vertebrates into print in a mad effort to gain priority over the other. A whole fauna of giant dinosaurs sprang into the scientific and public eye. Some of these, such as *Brontosaurus* and *Stegosaurus*, became synonymous with the very idea of dinosaurs. Many of the animals described by these rivals received two scientific names—a Cope name and a Marsh name—which had to be sorted out later by a less partisan generation.

Cope was a charming if quarrelsome and over-imaginative character. After a brief stint on the faculty of Haverford College, Cope used the fortune he had inherited from his father to set up as an independent scientist. He was a prodigious worker with a passion for describing new species. In all he was to produce 1,400 published works, which is an amazing total considering that he worked alone and that many of these publications were of substantial size and of real importance. Some of the dangers and discomforts endured by Cope and his parties are told by Sternberg in *Life of a Fossil Hunter*.

Cope's rival, Marsh, was a more secretive and far less appealing individual than Cope. Unlike Cope, Marsh was the manager of a large research group and very much a part of the scientific establishment of his day. Nonetheless, Marsh's personal adventures were no less amazing. Marsh, a Yale professor, led forays of students into the West during the early 1870s. Later, his collections came from parties of hired collectors who spent the entire year in the field. His activities in the West were not confined to science. In 1874, he carried to Washington the complaints of the Sioux Chief Red Cloud that Indian agents were cheating their charges. These accusations became a cause célèbre in the press which finally led to a full investigation and ultimately the resignations of the Commissioner of Indian Affairs and the Secretary of the Interior. Red Cloud visited Marsh in New Haven almost a decade later, an event recorded in a well-known photograph of the famous chief with the equally dignified Marsh. Ultimately, the feud between Cope and Marsh became one of the diverting public scandals of its day. Sternberg played no direct part in the feud, but in his book he helps illuminate Cope's character and the feel of the times.

Charles Sternberg was born in rural upstate New York in 1850. His father was a Lutheran minister and principal of a seminary. Charles grew up in a religious environment and was to retain a devoutly religious outlook all of his life. While he was growing up, his older brother George served as a medical officer in the Union Army. After the end of the war, George was posted as an Army surgeon at Fort Harker, near the town of Ellsworth, Kansas. When the elder Sternberg resigned from his seminary as the result of a dispute with the board of directors, George invited his family to join him in operating a ranch he had acquired. In 1867 the Sternbergs and eight of their children moved to the Kansas prairie. Kansas was then a hard frontier country, where Indians bitterly resisted the intrusion of the railroad into their hunting grounds and buffalo chips were still used as fuel to boil coffee. To Charles, who had already become interested in fossils as a child in New York, Kansas was to be the opening of a new world. The Cretaceous Dakota sandstone with its beautifully preserved fossils provided his start as a professional fossil collector. These Dakota fossils included tree leaves similar to willow, sassafras, and sycamore, as well as many extinct forms. The Niobrara chalk with remains of great Cretaceous marine reptiles, long-necked plesiosaurs with short bodies and great oar-like paddles, streamlined mososaurs with powerful jaws armed with formidable teeth, and twelve-foot sea turtles, became a life-long passion. To Sternberg these were not merely bones. As he tells it himself, he mentally reconstructed these animals and imagined their vibrant, if violent, lives in the old Kansas sea.

Sternberg's start as a professional collector began after his first (and only) year in college at the Kansas State Agricultural College at Manhattan. His attempt in 1876 to join a party collecting for O. C. Marsh failed, and in desperation he wrote to Cope to ask for money to buy ponies, a wagon, and camp outfit, and to hire a cook and driver so that he could collect fossils in western Kansas. Amazingly, Cope responded, and the consequence for Sternberg was the setting of the course of his life.

By 1909 when Sternberg wrote *Life of a Fossil Hunter*, he had collected for years for Cope and finally become an independent collector. As they grew up he incorporated his three sons, George, Charles, and Levi, into a fully professional team of fossil hunters. Their specimens include many of the spectacular dinosaur skeletons that grace the major museums of the world, including the American Museum of Natural

History, the British Museum, the Museum of Natural History in Paris, and the Senckenberg Museum in Frankfurt-am-Main. Their most important find was their least expected, a complete mummified duck bill dinosaur whose skeleton remains largely covered by the petrified skin. That specimen was discovered in Wyoming in 1908 by George and Levi while Sternberg and his other son Charlie went into town to replace depleted supplies. In the few days that the wagon was gone and food dwindled to a monotonous diet of potatoes, George and Levi excavated a skeleton they had found in a sandstone ledge. When the sandstone covering was removed, a cast of the skin was revealed. This remarkable fossil was sold to the American Museum of Natural History: it is exhibited on its back very much as it was when first exposed. Not only is the skin preserved, but also tendons and now fossilized tatters of flesh. The animal's head is twisted behind its back. The impression is of the immediate drama of death, and it powerfully evokes a very different living world 65 million years ago. The Sternbergs were to find other skeletons with skin impressions in later years in Canada.

Sternberg wrote *Life of a Fossil Hunter* when he was already nearly sixty, but his long career war far from over. In 1912, under an agreement with the Geological Survey of Canada to provide dinosaurs to Canadian and British museums, Sternberg and his sons began work in the rich fossil deposits exposed in the badlands of the Red Deer River valley of Alberta. This contract reflected the belated realization by the Canadians that the Red Deer River beds were being busily mined by an American Museum party under another unusual and indefatigable field paleontologist, Barnum Brown. The two competing parties worked for several years from houseboats on the river in a friendly rivalry. In the middle of the First World War, Charles Sternberg resigned because of a disagreement with the director of the Geological Survey, and returned to Kansas. His sons continued to collect in Alberta. One of their shipments consigned to the British Museum fell victim to German torpedoes when the freighter *Mount Temple* was sunk. But there were other fossils, including duckbills, the carnivorous tyranosaur *Gorgosaurus*, and a large armored dinosaur. These were offered to the British Museum in 1917, but they were not willing to repeat the expensive experiment of shipping dinosaurs through the U-boat infested Atlantic. The mounted skeletons are now exhibited at the American Museum in New York.

Meanwhile Charles Sternberg turned to collecting fossils in the Southwest for the University of California. He also wrote a second book entitled *Hunting Dinosaurs on Red Deer River, Alberta, Canada*. After his wife's death in 1939, he finally retired, living in Canada with Levi until his death at 93.

Sternberg's sons all built substantial careers in paleontology. Levi continued to collect for the Royal Ontario Museum in Toronto, where he helped build the outstanding collection of dinosaurs now on exhibit there. Charlie was affiliated with the National Museum of Canada, for which he not only collected, but acted as a laboratory scientist involved in the analysis and description of fossils as well. He played a role in creation of the Dinosaur National Park in Alberta, where exhibits of complete articulated dinosaur skeletons have been prepared for viewing outdoors, *in situ* in the ancient river valley deposits from which they have been exposed by erosion and then by careful excavation. George took part in the 1922 expedition to Patagonia sponsored by the Field Museum in Chicago, and then settled in Oakley, Kansas. He finally accepted a post at Fort Hays State College to develop the school's museum. Returning to the scene of Charles Sternberg's early triumphs, George collected outstanding marine vertebrates from the Niobrara chalk of Kansas. Among the prizes of those years is a huge slab displaying the perfectly preserved remains of the fourteen-foot-long fish, *Portheus*, containing its last meal, another complete six-foot-long fish. This particular slab now rests with other Sternberg fossils at the museum in Fort Hays State University, now called the Sternberg Museum.

Charles Sternberg has left us a remarkable cultural heritage. The life he records in his book offers a unique picture of a scientific freelancer in an age of increasing specialization and professionalization. His story is as much a part of the frontier as the more familiar tales of the West. It is true that he collected specimens for others to study, but he did it with a rare genius. Finding fossil vertebrates is not easy. Excavating them is an enormous labor of patience and attention to detail. Nor was Sternberg simply a collector. He fleshed out his creatures with a remarkably vivid imagination. The Kansas sea was real for him, under a sky filled with great soaring pterodactyls.

Many of Sternberg's specimens had considerable scientific importance, and indeed even the leaves he collected at the dawn of his career, which now reside in the Smithsonian and other museums such as the Museum für Naturkund in Berlin, are still used for research into the origins of flowering plants. He and the other turn-of-the-century collectors gathered immense collections of fossils that made possible the remarkable development of American paleontology and filled the exhibit halls of the great museums. These discoveries, and the spectacular museum displays built around them, made the American public aware of the age of the Earth and its extraordinary history. The dinosaur galleries of museums retain their popularity, and in more than one instance they have hooked impressionable children on careers in science.

As for the dinosaurs themselves, they are now enjoying a renaissance. Where dinosaurs once served as metaphors for anything powerful but sluggish and stupid and foredoomed to extinction, they are now visualized as having active metabolisms and exhibiting elaborate social behaviors. New questions have been raised about their evolutionary histories: they are the probable ancestors of birds. Were dinosaurs warm-blooded? Were any of them covered by feathers? Dinosaur nests are known, complete with fossil eggs and juveniles. Did the parent dinosaurs care for their young? The hypothesis that the extinction of the dinosaurs and a host of other Mesozoic creatures resulted from the impact of a large asteroid with the Earth has led to a revival in the study of extinction and of the role of catastrophic events in evolution.

But Sternberg has his own story to tell.

INTRODUCTION

By HENRY FAIRFIELD OSBORN,

President and Curator of Vertebrate Paleontology of the American Museum of Natural History, New York

OUR bookshelves contain the lives or narratives of adventure of many hunters of living game, but the life of a fossil hunter has never been written before. Both are in the closest touch with nature and, therefore, full of interest. The one is as full of adventure, excitement and depression, hope and failure, as the other, yet there is ever the great difference that the hunter of live game, thorough sportsman though he may be, is always bringing live animals nearer to death and extinction, whereas the fossil hunter is always seeking to bring extinct animals back to life. This revivification of the past, of the forms which once graced the forests and plains, and rivers and seas, is attended with as great fascination as the quest of live game, and to my mind is a still more honorable and noble pursuit.

The richness of the great American fossil fields,

which extend over the vast arid and semi-arid area of the West, scattered over both the great plains region and the great mountain region, has resulted in the creation of a distinctively American profession: that of fossil hunting. The fossil hunter must first of all be a scientific enthusiast. He must be willing to endure all kinds of hardships, to suffer cold in the early spring and the late autumn and early winter months, to suffer intense heat and the glare of the sun in summer months, and he must be prepared to drink alkali water, and in some regions to fight off the attack of the mosquito and other pests. He must be something of an engineer in order to be able to handle large masses of stone and transport them over roadless wastes of desert to the nearest shipping point; he must have a delicate and skilful touch to preserve the least fragments of bone when fractured; he must be content with very plain living, because the profession is seldom, if ever, remunerative, and he is almost invariably underpaid; he must find his chief reward and stimulus in the sense of discovery and in the despatching of specimens to museums which he has never seen for the benefit of a public which has little knowledge or appreciation of the self-sacrifices which the fossil hunter has made.

The fossil fields of America have fortunately attracted a number of such devoted explorers, and one of the pioneers on the honorable list is the author of this work, who by his untiring energy has contributed some of the finest specimens which now adorn the shelves and cases of many of the great museums of America and Europe.

Although special explorations have been described, sometimes in considerable detail, this is the first time that the "life of a fossil hunter" has been written, and it is fitting that it comes from the pen of the oldest living representative of this distinctively American profession. The name of Charles H. Sternberg is attached to discoveries in many parts of the West; discoveries which have formed distinct contributions to science, to the advance of paleontology, to our knowledge of the wonderful ancient life of North America. His is a career full of adventure, of self-sacrifice, worthy of lasting record and recognition by all lovers of nature.

PREFACE

I WISH to call the attention of the reader of my story "The Life of a Fossil Hunter" to the fact that I am under obligations especially to Prof. Henry Fairfield Osborn, President and Curator of Paleontology of the American Museum of Natural History in New York. He has supplied me with many of the most beautiful of the illustrations that illumine these pages and has assisted the work in many ways.

I would also express my gratitude to Miss Margaret Wagenalls of New York, who edited the manuscript; to Prof. Dunlap of the Kansas State University, for his kindly criticisms; and to Dr. W. K. Gregory, Lecturer on Zoology at Columbia University, whose untiring efforts have brought the book to its present form.

I hope it may awaken a wide interest in the study of ancient life, and I thank my friends everywhere who are contributing to that end.

CHARLES H. STERNBERG.

LAWRENCE, KANSAS,
January, 1909.

THE LIFE OF A FOSSIL HUNTER

THE LIFE OF A FOSSIL HUNTER

CHAPTER I

EARLY DAYS AND WORK IN THE DAKOTA GROUP OF THE CRETACEOUS

DO not remember when I first began collecting fossils, but I have always loved nature.

Fifteen years of my early life were spent in Otsego County, New York, at dear old Hartwick Seminary, where my father, the Rev. Dr. Levi Sternberg, was principal for fourteen years, and my grandfather, Dr. George B. Miller, a much-loved, devout man, professor of theology for thirty-five. The lovely valley of the Susquehanna, in which it stands, lies five miles below Cooperstown, the birthplace of the Walter Scott of America, James Fenimore Cooper, and my boyhood was spent among scenes which he has made famous. Often my companions and I have gone picnicking on Otsego Lake, shouting to call up the echo, and spreading our tablecloth on shore beneath the very tree from which the catamount was once about to spring upon terrified Elizabeth Temple.

My greatest pleasure in those early days and best, was to live with a darling cousin in the woods. There among the majestic trees,—maples, hickories, pines, and hemlocks,—we used to build sylvan retreats, weaving willow twigs in and out among the poles which I cut for supports; and there, to those great trees, I delivered my boy orations. We delighted also to visit and explore Moss Pond, a body of water on top of the hills across the river, surrounded entirely by sponge moss. We could "teeter" across the moss to a log that gave us support, and catch blind bullheads, or eat our lunch in the cool, dense hemlock woods that surrounded the water, where the heavy branches, intertwined like mighty arms, shut away the light, so that even at midday the sun could barely pierce their shadows.

How I loved flowers! I carried to my mother the first crocus bloom that showed its head above the melting snow, the trailing arbutus, and the tender foliage of the wintergreen. Later in the season I gathered for her the yellow cowslip and fragrant water-lily; and when autumn frosts had tinged the leaves with crimson and gold I filled her arms with a glorious wealth of color.

Even in those early days I used to cut out shells

from the limestone strata of the region with whatever tools were at hand, but they were admired chiefly as examples of the wonderful power of running water to carve rocks into the semblance of shells. Or if one of the more observant remarked that these shells looked very much as if they had been alive once, the only theory that would account for their presence and yet sustain the belief that the world was only six thousand years old, was that the Almighty, who created the rocks, could easily, at the same time, have created the ancient plants and animals as fossils, just as they were found.

I remember a rich find I made in the garret of an uncle in Ames, New York,—a cradle filled with fossil shells and crystals of quartz. They had been collected by my uncle's brother, who, fortunately, as my uncle said, had died early, before bringing disgrace upon the family by wasting his time wandering over the hills and gathering stones. All the large specimens he had collected had been thrown away, and the smaller ones in the old cradle had long been forgotten. I was welcome to all my uncle's buggy could carry when he took me home, and I can never forget the joy of going over that material again and again, selecting the specimens that appealed most to my sense of the beautiful and the wonderful. I labeled them all "From Uncle James," and it greatly astonished a dear aunt of mine, to whom I gave them some years later when we moved West, to find in the collection a lot of baculites, labeled "Worms from Uncle James."

When I was ten years old, I met with an accident from which I have never completely recovered. I remember the wild chase I was making after an older boy, over the hay-mows and piles of shocked grain in my father's barn. On the floor below, an old-fashioned thresher, one of the first of its kind, was making an ear-splitting noise, while outside the two horses, hitched to an inclined plane, climbed incessantly, but never reached the top.

The boy climbed a shock of oats on the scaffold in the peak of the barn, and "Charley-boy," as my mother called me, following him, slipped through a hole in the top of the ladder which had been covered by the settling oats, and fell twenty feet to the floor below. The older boy climbed swiftly down and carried me home insensible to my mother.

Our family physician thought that only a sprain was the result, and bandaged the injured limb; but, as a matter of fact, the fibula of the left leg had been dislocated, so that there was much suffering and a little crippled boy going about among the hills on crutches.

The leg never grew quite strong again, and some years later gave me a good deal of trouble. In 1872 I was in charge of a ranch in Kansas, and during November of that year a great sleet storm covered the whole central part of the state. In order to water my cattle, which were scattered over a range of several thousand acres on Elm Creek, I was obliged to follow around small bands of them to their accustomed watering-places and cut the ice for them. The water that splashed over my clothing froze solid, and the result was that inflammatory rheumatism settled in the lame leg. I sat in a leathern chair all winter close to a boxwood stove, tended by my dear mother, who never left me day or night.

When the inflammation subsided, the knee joint had become ankylosed, and in order to avoid going on crutches all my life, I lay in the hospital at Fort Riley for three months, all alone in a great ward, and had the limb straightened by a special machine. So skilfully did the army surgeon do this work that I threw away crutches and cane, and, although the leg has always been stiff, I have since walked thousands of miles among the fossiliferous beds in the desolate fields of the West.

In 1865, when I was fifteen years old, my father accepted the principalship of the Iowa Lutheran College at Albion, Marshall County, and the broken hill country of my boyhood days was replaced by the plains and water courses of the Middle West.

Two years later my twin brother and I emigrated

to an older brother's ranch in Ellsworth County, Kansas, two and a half miles south of Fort Harker, now known as Kanopolis. This post was at that time the terminus of the Kansas Division of the Union Pacific, and almost daily train-load after train-load of prairie schooners, drawn by oxen, burros, or mules, pulled out from it over the old Butterfield and Santa Fé trails, the one leading up the Smoky Hill, the other through the valley of the Arkansas to Denver and the Southwest.

In spring great herds of buffalo followed the tender grass northward, returning to the South in the fall; and one bright day my brother and I started out on our first buffalo hunt. Driving a team of Indian ponies hitched to a light spring wagon, we soon left the few settlements behind, and reached the level prairie to the southwest, near old Fort Zaro, a deserted one-company post on the Santa Fé Trail. At this time it had been appropriated by a cattleman who had a small herd grazing in the vicinity.

When within a few miles of this post, we saw a large herd of buffalo lying down a mile away. It was no easy matter to crawl toward them over the plain, pushing myself along without raising my body above the short grass, but after strenuous efforts I got within shooting distance without disturbing them, and was resting for a shot, when the rancher rode through the herd and sent them all off at a lope. Much angered and almost tempted to turn my gun on the man, I returned to the wagon, and we drove on across country that had been cropped as if by a great herd of sheep by the thousands of buffalo that had passed that way on their journey south.

Anxious to find picketing-ground and water, we reached the Arkansas River, where in a swale covered with grass and willows were paths cut by the buffalo. I lay down in one of these, and bringing my gun to my shoulder, was just drawing bead, when a large animal rushed across my line of vision at right angles to the trail. I pulled the trigger, and down went the brown mass in a heap on the ground.

Swinging my gun above my head, I rushed forward shouting, "I've killed a buffalo!"—to find that I had shot a Texas cow. Terrified at the thought of its owner's anger, we rushed back to the wagon, and, whipping up the ponies, sped away as if the furies were after us. But cooler second thoughts led us to the conclusion that the cow had come north with the buffalo, and was as much our prey as the buffalo themselves.

Just before sunset we reached a part of the country through which the buffalo had not passed, where a rich carpet of grass, covering all the plain, offered plenty of food for our tired ponies. Here we were delighted to find, standing in a ravine, an old bull buffalo, which had been driven out of the herd to die. Concealing ourselves behind the carcass of a cow, we opened fire upon him from our Spencer carbines, and continued to riddle his poor old body with leaden slugs until his struggles ceased. Even then, when he had lain down to rise no more, we crawled up behind him and threw stones at him, to make sure that he was dead. We found his flesh too tough for food; but it was an exciting event to us two boys to kill this massive beast, in earlier days perhaps the leader of the herd.

In this connection I might tell of a chase I had several years later, while living on a ranch in eastern Ellsworth County. I saw a huge buffalo bull come loping along from the hills, headed for a section of land that was inclosed by a wire fence. On the other side of this section there was a piece of timber-land, and fearing that if he got into the dense timber I should lose him, I rode after him at the top of my speed.

When his lowered head struck the wire fence it flew up like a spring gate and immediately closed down behind him. In order to follow, I had either to cut the wire or go out of my way to a gate half a mile to the south. I decided on the latter course, and applied quirt and spur to my horse, but upon reaching the gate, discovered my escaping quarry already halfway across the section. I got just near enough to put a bullet into his rump as he passed through the fence on the other side, and disappeared in the dense woods beyond.

In my excitement I shouted to my pony, and, dismounting and standing on the wire to hold it down, yelled at him to come across. But a sudden fit of obstinacy had seized him, and he would not come. I had to let the fence up while I thrashed him, and then as soon as I got it under my feet again, he pulled back as before. We repeated this performance until I was exhausted and gave up the struggle.

But upon casting a look of despair in the direction of the vanished buffalo, I was both astonished and ashamed to see him standing under an elm tree not ten feet away, covered up all except his eyes by a great wild grapevine, and gazing in mute astonishment at the struggle between Nimrod and his pony. I have always regretted that I took advantage of the confidence he placed in me, for as soon as I could control my jumping nerves, I shot the noble beast behind the shoulder, and he fell.

I saw my last herd of buffalo in Scott County, Kansas, in 1877. Antelope, however, continued to be abundant as late as 1884, and only two years ago I saw a couple of them among some cattle near Monument Rocks, in Gove County.

In camp, during those early days, we were rarely out of antelope meat, and even now my mouth waters at the thought of the delicious tenderloin, soaked first in salt water to season it and remove the blood, then covered with cracker dust, and fried in a skillet of boiling lard. In those days a hind quarter could be hung up under the wagon in the hottest part of summer, and not spoil. The wind hermetically sealed it, and there were no blow-flies then. The early settlers of a new country bring with them, and protect, their enemies, and destroy their friends, the skunks, badgers, wildcats, and coyotes, as well as hawks, eagles, and snakes, because they kill a chicken or two as a change from their usual diet of prairie dogs and rabbits.

In those pioneer days the Kiowas, Cheyennes, Arapahoes, and other Indian tribes made constant inroads upon the venturesome settler who, following the advice of Horace Greeley, had come West to grow up with the country.

I remember when old Santante, a chief of the Kiowas, came to the post in a government ambulance, which he had captured on one of his raids. In time of peace, the Indians belong to the Interior Department of the government, so that all the officer in command at the fort could do was to extend the old chief the courtesy of the army and care of himself and team. Once, at the old stone sutler's store, I heard him remark, after he had filled himself well with whisky, "All the property on the Smoky Hill is mine. I want it, and then I want hair."

He got both the following year.

In July, 1867, owing to the fear of an Indian outrage, General A. J. Smith gave us at the ranch a guard of ten colored soldiers under a colored sergeant, and all the settlers gathered in the stockade, a structure about twenty feet long and fourteen wide, built by setting a row of cottonwood logs in a trench and roofing them over with split logs, brush, and earth. During the height of the excitement, the women and children slept on one side of the building in a long bed on the floor, and the men on the other side.

The night of the third of July was so sultry that I concluded to sleep outside on a hay-covered shed. At the first streak of dawn I was awakened by the report of a Winchester, and, springing up, heard the sergeant call to his men, who were scattered in rifle pits around the building, to fall in line.

As soon as he had them lined up, he ordered them to fire across the river in the direction of some cottonwoods, to which a band of Indians had retreated. The whites came forward with guns in their hands and offered to join in the fight, but the sergeant commanded: "Let the citizens keep in the rear." This, indeed, they were very willing to do when the order was given, "Fire at will!" and the soldiers began sending leaden balls whizzing through the air in every conceivable arc, but never in a straight line, toward the enemy, who were supposed to be lying on the ground.

As soon as it was light my brother and I explored the river and found a place where seven braves, in their moccasined feet, had run across a wet sandbar in the direction of the cottonwoods, as the sergeant had said. Their pony trails could be easily seen in the high, wet grass.

The party in the stockade were not reassured to hear the tramp of a large body of horsemen, especially as the soldiers had fired away all their ammunition; but the welcome clank of sabers and jingle of spurs laid their fears to rest, and soon a couple of troops of cavalry, with an officer in command, rode up through the gloom.

After the sergeant had been severely reprimanded for wasting his ammunition, the scout Wild Bill was ordered to explore the country for Indian signs. But, although the tracks could not have been plainer, his report was so reassuring that the whole command returned to the Fort.

Some hours later I spied this famous scout at the sutler's store, his chair tilted back against the stone wall, his two ivory-mounted revolvers dangling at his belt, the target of all eyes among the garrison loafers. As I came up this gallant called out, "Well, Sternberg, your boys were pretty well frightened this morning by some buffalo that came down to water."

"Buffalo!" I said; "that trail was made by our old cows two weeks ago."

Later the general in command told me that they had prepared for a big hop at the Fort on the night of the fourth, and that Bill did not report the Indian tracks because he did not want to be sent off on a long scout just then.

In the unsettled state of the country at this time there were other dangers to be guarded against beside that of Indians, as I learned to my cost.

As a boy of seventeen, it was my duty on the ranch to haul milk, butter, eggs, and vegetables to Fort Harker for sale. I cared for my pony myself, and in order to get the milk and other food to the Fort in time for the soldiers' five-o'clock breakfast, I had to go without my own. One day I had a number of bills to collect from the officers, but as I was unusually tired, and the officers were not out of bed when I called, I put the bills in my inside pocket and started home.

As was my custom, after leaving the garrison I lay down on the wagon-seat and went to sleep, letting my faithful horse carry me home of his own accord. I have no recollection of what happened afterwards, but when I reached the ranch my brothers found me sitting up in the wagon moaning and swinging my arms, with the blood flowing from a slung-shot wound in my forehead. I had been struck down in my sleep and robbed of all the money I had on my person, as it happened only about five dollars.

Providentially our nearest neighbor, D. B. Long, was a retired hospital steward, and the post surgeon, Dr. B. F. Fryer, who was sent for immediately, was just ready to drive to town with his team of fleet little black ponies. He reached the ranch in an incredibly short time, and, although respiration had ceased, those two faithful men kept up artificial respiration for hours. My oldest brother, Dr. Sternberg, for years Surgeon-General of the Army, was also sent for, and I found him lying on a mattress by my side when I regained consciousness two weeks later.

I might tell also of the ruffians who at one time held Ellsworth City in a grip of iron, and how, until they killed each other off or moved further west with the railroad, the dead-cart used to pass down the street every morning to pick up the bodies of those who had been killed in the saloons the night before, and thrown out on the pavement to be hauled away.

But, although I should like to recall more of the incidents connected with the opening up of a new country, time presses, and I must pass on to an account of my work as a fossil hunter.

I had not been long in this part of the country before I found that the neighboring hills, topped with red sandstone, contained, in isolated places, from a few feet to a mile in diameter and scattered through a wide expanse of country, the impressions of leaves like those of our existing forests.

The rocks consisted of red, white, and brown sandstone, with interlaid beds of variously-colored clays; while here and there, scattered through the formation, were vast concretions of very hard flint-like sandstone, often standing on softer rocks that had been weathered away into columns, the whole giving the effect of giant mushrooms, as seen in the cuts (Figs. 1-3).

This formation, resting unconformably on the upper carboniferous rocks, belongs to the Dakota Group of the Cretaceous Period. The sedimentary rocks were laid down during the Cretaceous Period, the closing period of the "Age of Reptiles," in a great ocean, whose shore line enters Kansas at the mouth of Cow Creek on the Arkansas River, and extending in a northwesterly direction in the vicinity of Beatrice, Nebraska, touches Iowa, and passes on to Greenland.

I was carried away at this time by the thoughts that had been surging through the hearts of men since Darwin bade them turn to nature for the answers to their problems concerning the plants and animals of this earth.

How often in imagination I have rolled back the years and pictured central Kansas, now raised two thousand feet above sea level, as a group of islands scattered about in a semi-tropical sea! There are no frosts and few insect pests to mar the foliage of the great forests that grow along its shores, and the ripe leaves fall gently into the sand, to be covered up by the incoming tide and to form impressions and counterparts of themselves as perfect as if a Divine hand had stamped them in yielding wax.

Go back with me, dear reader, and see the treeless plains of to-day covered with forests. Here rises the stately column of a redwood; there a magnolia opens its fragrant blossoms; and yonder stands a fig tree. There is no human hand to gather its luscious fruit, but we can imagine that the Creator walked among the trees in the cool of the evening, inhaling the incense wafted to Him as a thank-offering for their being. All His works magnify Him. The cinnamon sends forth its perfume beside the sassafras; linden and birch, sweet gum and persimmon, wild cherry and poplar mingle with each other. The five-lobed sarsaparilla vine encircles the

FIG. 1.—ROCKS OF LARAMIE BEDS ON SOUTH SCHNEIDER CREEK, CONVERSE COUNTY, WYOMING.

FIG. 2.—WEATHERED ROCKS AND LARAMIE BEDS NEAR SOUTH SCHNEIDER CREEK.

FIG. 3.—MUSHROOM-LIKE CONCRETION KNOWN AS PULPIT ROCK.
Elm Creek, Kansas, near Sternberg's ranch. (From Trans. Kan. Acad. Sci.)

tree-trunks, and in the shade grows a pretty fern. Many other beautiful plant forms grace the landscape, but the glorious picture is only for him who gathers the remains of these forests, and by the power of his imagination puts life into them; for it is some five million years, according to the great Dana of my childhood days, since the trees of this Kansas forest lifted their mighty trunks to the sun.

At the age of seventeen, therefore, I made up my mind what part I should play in life, and determined that whatever it might cost me in privation, danger, and solitude, I would make it my business to collect facts from the crust of the earth; that thus men might learn more of "the introduction and succession of life on our earth."

My father was unable to see the practical side of the work. He told me that if I had been a rich man's son, it would doubtless be an enjoyable way of passing my time, but as I should have to earn a living, I ought to turn to some other business. I say here, however, lest I forget it, that, although my struggle for a livelihood has been hard, often, indeed, bitter, I have always been financially better off as a collector than when I have wasted, speaking from the point of view of science, some of the most precious days of my life attempting to make money by farming or in some other business, so that I might live at home and avoid the hardships and exposures of camp life.

With collecting-bag over my shoulder and pick in hand, I wandered over the hills of Ellsworth County. If I chanced upon a locality rich in fossil leaves, thrilled with a joy that knows no comparison, I walked on air as I carried my trophies home; while if night overtook me with an empty bag, I could scarcely drag my weary limbs along.

Among the rich localities that I discovered was one which I called "Sassafras Hollow," because of the countless sassafras leaves I quarried there. It it situated about a mile southeast of the schoolhouse on Thompson Creek, in the Hudson brothers' neighborhood, and lies at the head of a narrow ravine in a ledge of sandstone, with a spring beneath. Here too, the noted paleobotanist, Dr. Leo Lesquereux, collected fossils in 1872, securing among other specimens a large, beautiful leaf which he named in my honor "*Protophyllum sternbergii.*"

I have a vivid recollection of the discovery of another locality. One night I dreamed that I was on the river, where the Smoky Hill cuts into its northern bank, three miles southeast of Fort Harker. A perpendicular face in the colored clay impinges on the stream, and just below this cliff is the mouth of a shallow ravine that heads in the prairie half a mile above.

In my dream, I walked up this ravine and was at once attracted by a large cone-shaped hill, separated from a knoll to the south by a lateral ravine. On either slope were many chunks of rock, which the frost had loosened from the ledges above. The spaces left vacant in these rocks by the decayed leaves had accumulated moisture, and this moisture, when it froze, had had enough expansive power to split the rock apart and display the impressions of the leaves.

Other masses of rock had broken in such a way that the spaces once filled by the midribs and stems of the leaves admitted grass roots; and their rootlets, seeking the tiny channels left by the ribs and veins of the leaves, had, with the power of growing plants, opened the doors of these prisoners, shut up in the heart of the rock for millions of years.

I went to the place and found everything just as it had been in my dream.

Two of the largest leaves known to the Dakota Group were taken from this place. One, a great three-lobed leaf, the stem passing through an ear-like projection at its base, Dr. Lesquereux called *Aspidophyllum trilobatum;* the other, equally large, —over a foot in diameter,—and three-lobed too, but indented with large teeth, he called *Sassafras dissectum* (Fig. 4).

I believe I am the only fossil hunter who has

collected from this locality. Probably my eyes saw the specimens while I was chasing an antelope or stray cow and too much occupied with the work in hand to take note of them consciously, until they were revealed to me by the dream, the only one in my experience that ever came true. I tell this story to show how deeply I was interested in these fossils.

My first collection, or rather the cream of it, was sent to Professor Spencer F. Baird, of the Smithsonian Institution. The following is the letter which I received from him:

SMITHSONIAN INSTITUTION,

Washington, June 8, 1870.

Dear Sir:—We are duly in receipt of your letter of May 28th, announcing the transmission of the fossil plants collected by your brother and yourself, and shall look forward with much interest to their arrival. As soon as possible after they reach us, we shall submit them to competent scientific investigation, and report to you the result.

Very respectfully yours, etc.,

SPENCER F. BAIRD,

Assistant Secretary in Charge.

There was no money in fossils at that early day, but I prized more highly than money the promise in the letter that my specimens would be studied by competent authority, and that I should receive credit for my discoveries.

FIG. 4.—FOSSIL LEAVES OF *Sassafras dissectum.*
(After Lesquereux.)

FIG. 5.—*a*, UNOPENED LEAF NODULE; *b*, NODULE OPENED TO SHOW FOSSIL LEAF; *c*, *d*, *e*, *f*, VARIOUS FORMS OF FOSSIL LEAVES.

The specimens were sent to Dr. John Strong Newberry, professor in Columbia University and State Geologist of Ohio. He did not find opportunity at that time to publish the results, but long years afterwards, in 1898, I received from Dr. Arthur Hollick a copy of "Later Flora of North America," a posthumous work of Dr. Newberry's. Turning instantly to the magnificent plates, I recognized some of my early specimens, the first I ever collected that were of value to science.

Although, owing to the long delay in publication, I lost credit for them, and the duplicates which I had given to a friend had been used by Lesquereux to illustrate some new species accredited to that friend instead of to their rightful discoverer, Dr. Newberry kindly acknowledged my work on p. 133 of his book, where he says: "The leaf figured on Plate X and that represented on Plate XI were included in a collection made by Charles H. Sternberg, and Lesquereux has done only justice to him by attaching his name to the finest species contained in the large collection of fossil plants he made there," that is, at Sassafras Hollow.

In 1872, just before Lesquereux's great work, "The Cretaceous Flora," appeared, I learned that the famous botanist was a guest of Lieutenant Benteen, the commander of Fort Harker. Fortunately, I had retained rough sketches of the first specimens I had sent to the Smithsonian Institution. So with these I started for the Post, where I found a reception in progress in honor of the noted guest.

I was introduced to the venerable botanist by his own son, who spoke to him in French, as he was almost deaf. When I displayed my sketches, he took me to one side, and in a corner of the room I told him the story of my discoveries. His eyes shone when he examined the drawings. "This is a new species," he said, "and this, and this. Here is one described and illustrated from poorer material."

I do not remember how long we talked. I only know that the golden moments sped by all too rapidly; and from that hour until his death in 1889 we were in constant correspondence.

After this all my collections were sent to him for description. Over four hundred species of plants like those of our existing forests along the Mexican Gulf, some beautiful vines, a few ferns, and even the fruit of a fig, and a magnolia flower petal, the only petal so far found in the coarse sandstone of the Dakota Group, have rewarded my earnest efforts. The fragrance of this lovely flower seems wafted down to us through the myriads of ages since it bloomed.

Dr. Arthur Hollick, in his paper, "A Fossil Petal

and a Fruit from the Cretaceous (Dakota Group) of Kansas," in Contributions from the New York Botanical Garden, No. 31, says, on page 102: "Included in a collection of fossil-plant remains from the Cretaceous (Dakota Group) of Kansas, recently obtained by the New York Botanical Garden from Charles H. Sternberg of Lawrence, Kansas, are two exceedingly interesting specimens,—one representing a large petal, the other a fleshy fruit. Petals are exceedingly rare, and I am not acquainted with any published figure of anything of the kind which can compare with ours in regard to either size or satisfactory condition of preservation."

Of the fig, the Doctor remarks: "The fruit is plainly that of a fig, and, although some twenty-three species of *Ficus* have been described from the Dakota Group, they were based upon leaf impressions. This fossil has every appearance of many dried herbarium specimens, and it is evident that it must have possessed considerable consistency in order to retain its original shape, as it has done to a certain extent, under the pressure to which it must have been subjected."

In 1888 I sent over three thousand leaf impressions from the Dakota sandstone to Dr. Lesquereux, and he selected from them over three hundred and fifty typical specimens, many of them new, for the National Museum. Hundreds of others, identified by him, were afterwards purchased by R. D. Lacoe, of Pittston, Pa., and presented to the Museum.

So feeble had the great botanist become in these last years of his life, that friends passed before his failing eyes the trays containing these great collections.

In my estimation, America can show no life more unselfishly devoted to science than that of Lesquereux, probably the most scholarly and conscientious botanist of his day. He once wrote me that he received a salary of five dollars a day from the U. S. Geological Survey, and out of this he had to pay his artist. He labored with unfailing enthusiasm to complete his monumental work, " The Flora of the Dakota Group," but by the irony of fate, he never saw his beloved book in print. It was published by the Government five years after his death, under the able editorship of Dr. F. H. Knowlton.

He passed away at the age of eighty-three.

" Born in the heart of Switzerland's mountain grandeur," he once said, " my associations have been almost all of a scientific nature. I have lived with nature,—the rocks, the trees, the flowers. They know me, I know them. Everything else is dead to me."

It was my good fortune to be in constant correspondence with Lesquereux, and his letters, which I need not say I prize highly, have done more, per-

Columbus O 14th April 75

Mr Chs Sternberg Fort Harker

My dear Sir

I much approve of your purpose of studying medicine. Your taste for natural history will help you much and encourage you. But allow me still to say, to you as a friend would do, that you can not expect to become useful to others and to yourself in science except by hard work, pursued with patience and a fixed purpose. Science is a high mountain. To go up to its top or at least high enough to gain free atmosphere and wide horizon necessitates hard climbing, through brushes thickets rocks etc. Those who from the beginning look around for commodious and soft paths merely enter the gloom of the woods at the base. They are seen from nobody and see nothing but undistinct forms and because there horizon is thus limited to darkness they think there is nothing else and nothing more to see from high above toward the top of the mountain. Moreover there is not a true hard step in science or in life which does not give its reward in one way or another While we have not a single moment of laziness of unmerited comfortable rest; which does not bring us some kind of disappointment and has not to be paid by a little more trouble and work

Yours very truly,

L. Lesquereux

haps, than any other thing to fix my determination that, come what might, I would be a fossil hunter and add my quota to human knowledge. The letter here reproduced has been as a lodestar to lead me on past all discouragements in the path which as a boy of seventeen I set out to follow. May it shed light upon the life of some other struggler!

In 1897, not having the means to go into the vertebrate fields of western Kansas, I spent three months in the Dakota Group, although I knew that I had already supplied most of the museums of the world with examples of its flora, and that there was little interest in or demand for the leaves.

I secured over three thousand leaves, however, and paid first-class freight on them to my home at Lawrence. Then I hauled them out to my little twenty-acre farm, four miles southeast of town, and pitched my 9 x 9 wall-tent for a workshop, flooring it and putting up a stove. There I worked from November to May, standing on my feet on an average of fourteen hours a day, with my face to the opening of the tent for light, and my back to the stove. At night I worked over a coal-oil lamp.

With a chisel-edged hammer weighing two ounces, I trimmed off the rough stone from the margin of the nodules, as illustrated in the woodcuts by Christian Weber of New York (Fig. 5, *c, d, e,* and *f*), a labor of love on his part, for which I am deeply grateful. I smoothed down the rock with emery-stone also, and with a No. 1 needle pried away the stone from the petioles, leaving the impression as if it were the leaf itself standing up in bold relief, thus bringing out all its beauty. One of my neighbors, after examining the prepared specimens, remarked, "You must have taken a long time to carve those things. Why, they look just like leaves!"

When no more loving labor could be bestowed on them without risk of injuring the specimens, I laid them away in trays, to be numbered and identified. I knew that some authorities demanded the specimens in payment for the labor of identification, and as I had to make a living out of my work, this would never do for me. So after Lesquereux's death I undertook the work of identification myself, although I confess it hurt my conscience, as I had never had the training of a botanical authority. I was greatly relieved, therefore, when, after selling two hundred and fifty specimens to the New York Botanical Gardens, I asked Dr. Arthur Hollick whether my identifications were correct, to receive the answer that upon a casual examination he could find no reason to make any changes in my names. I was certainly much encouraged by such words from this eminent authority in fossil botany.

To return to my great collection from the Dakota

Group, I spent nine months of incessant labor upon it, and my readers may be surprised to learn that I was delighted when Professor Macbride, of the University of Iowa, purchased it for the munificent sum of three hundred and fifty dollars, the price I put upon it. My delight was even greater when I received the following letter, which is now and was then more highly prized than the check which it enclosed.

STATE UNIVERSITY OF IOWA.
BOTANY.
Iowa City, Iowa, May 1, 1898.

DEAR MR. STERNBERG:

The boxes are all safely here. We have at present no place for the display of the specimens, but have opened the first three cases and are delighted with the beauty of the material. I hope next year to have a case for fossil plants, when I shall certainly make a display of these beautiful leaves, and quote you as collector. I should think the National Museum would give you employment all the time.

I trust you may have a pleasant and profitable summer, and if in future I can in any way serve you, kindly advise me.

Very truly yours,
THOMAS K. MACBRIDE.

This small sum enabled me to go with my son George into the chalk of Kansas, where we discovered the splendid specimen of a mosasaur, now in the museum of Iowa University. But for the timely assistance given me when I most needed help, it is doubtful whether Iowa would have secured this treasure. My months of patient labor on the leaves had convinced the authorities that my work on the mosasaur would be faithfully done.

Before closing this account of my work in the Dakota Group, I should like to say a few words about the manner in which the nodules are formed around leaf impressions, a subject of which I have made a careful study during years of exploration. The illustrations (Fig. 5, *a* and *b*) show the nodules before they are opened, and the open specimens before they have been trimmed, as in the other cuts.

The mother rock, or matrix, as it is called, from which these concretions come, is quite soft and easily disintegrates into yellowish sand under the influences of the weather. Through this yellowish sandstone are scattered countless leaf impressions and their counterparts, but on account of the softness of the matrix it is impossible to work out any leaves from the inside of the rock masses, and we should lose them altogether were it not for the following natural process:

Falling from the trees that grew along the shore of the Cretaceous Ocean, these leaves were covered with sand by the incoming tide. Some, falling stem first, were turned over into a U-shape; others are found lying flat, and others again at various angles. The sand, accumulating through the years, finally became consolidated, and, being in course of time exposed to the air, began to "weather." In the meantime the iron coloring matter of the vegetation had been dissolved out by the water and distributed through the rock mass. As the rock weathers away, the leaf impressions are hardened by the iron that has been dissolved out of the sandy mass by water holding acids in solution.

As the soft rock about them continues to wear away, the nodules begin to appear above the surface, at first only as bumps slightly elevated above the surrounding rock, but in time as complete concretions, with the form of the leaves imprisoned within, which are left standing on pedestals no thicker than a lead pencil.

Then the first storm of rain or hail breaks them from their moorings; they become independent, are reduced in size, and constantly hardened, so that often a nodule is almost pure iron ore a fraction of an inch in thickness.

So the process goes on and will continue until all the leaves within the parent rock have been protected by an iron envelope; and it is this natural process alone which can save these beautiful impressions from falling to pieces when the sand is freed from the rock by disintegration.

The locality from which I collected these specimens I have named the *Betulites* locality, on account of the abundance of birch leaves of many varieties which have been found there. It was discovered by the late Judge E. P. West, collector for the University of Kansas, and Professor Lesquereux honored him by calling one species *Betulites westii.* He made a wonderful collection of Dakota leaves for the University, many of them new to science. The locality is about a mile in length and tops the highest hills in Ellsworth County.

I have no record of the thousands of fossil leaves I have collected from the sandstone of central Kansas. I have never kept a single specimen for myself, although I love them dearly, and it has often been hard to give them up. But the object of my life has been to advance human knowledge, and that could not be accomplished if I kept my best specimens to gratify myself. They had to go, and they went, often for less than they cost me in labor and expense, into the hands of those who could give authoritative knowledge of them to the world, and preserve them in great museums for the benefit of all.

One thing I have demanded as my right, in my opinion an inalienable right, although I am sorry to say that there are those who have denied it to me,—I demand that my name appear as collector on all the material which I have gathered from the rocks of the earth.

I might have sold to showmen or dealers; in fact I have the assurance of one of the largest dealers in America that I made a great mistake in selling directly to museums instead of through him. If I had done as he advised, the thousands of fossils I have collected would have cost the museums fifty per cent. more than they have, and my work would have been measured by the money these dealers would have been pleased to allow me, and I should never have been known as one of those who have devoted their lives to the advancement of paleontology.

CHAPTER II

MY FIRST EXPEDITION TO THE KANSAS CHALK, 1876

SPENT the winter of 1875 and '76 as a student at the Kansas State Agricultural College.

Here a party was gathered to explore western Kansas for fossils, under the leadership of Professor B. F. Mudge, the enthusiastic state geologist and a popular professor of the college. The expedition was to be made under the auspices of Professor O. C. Marsh, of Yale College, whose efforts have secured for that institution the largest collection perhaps in the world of American fossil vertebrates.

I made every effort in my power to secure a place in the party, but failed, as it was full when I applied. It has always been hard, however, for me to give up what I have determined to accomplish; so, although almost with despair, I turned for help to Professor E. D. Cope, of Philadelphia, who was becoming so well known that a report of his fame had reached me at Manhattan.

I put my soul into the letter I wrote him, for this was my last chance. I told him of my love for science, and of my earnest longing to enter the chalk of western Kansas and make a collection of its wonderful fossils, no matter what it might cost me in discomfort and danger. I said, however, that I was too poor to go at my own expense, and asked him to send me three hundred dollars to buy a team of ponies, a wagon, and a camp outfit, and to hire a cook and driver. I sent no recommendations from well-known men as to my honesty or executive ability, mentioning only my work in the Dakota Group.

I was in a terrible state of suspense when I had despatched the letter, but, fortunately, the Professor responded promptly, and when I opened the envelope, a draft for three hundred dollars fell at my feet. The note which accompanied it said: "I like the style of your letter. Enclose draft. Go to work," or words to the same effect.

That letter bound me to Cope for four long years, and enabled me to endure immeasurable hardships and privations in the barren fossil fields of the West; and it has always been one of the joys of my life to have known intimately in field and shop the greatest naturalist America has produced.

As soon as the frost was out of the ground, having secured a team of ponies and a boy to drive them, I left Manhattan and drove out to Buffalo Park, where one of my brothers was the agent. The only house, beside the small station building, was that occupied by the section men. Great piles of buffalo bones along the railroad at every station testified to the countless numbers of the animals slain by the white man in his craze for pleasure and money. A buffalo hide was worth at that time about a dollar and a quarter.

Here at Buffalo I had my headquarters for many years. A great windmill and a well of pure water, a hundred and twenty feet deep, made it a Mecca for us fossil hunters after two weeks of strong alkali water. At this well Professor Mudge's party and my own used to meet in peace after our fierce rivalry in the field as collectors for our respective paleontologists, Marsh and Cope.

What vivid memories I have of that first expedition!—memories of countless hardships and splendid results. I explored all the exposures of chalk from the mouth of Hackberry Creek, in the eastern part of Gove County, to Fort Wallace, on the south fork of the Smoky Hill, a distance of a hundred miles, as well as the region along the north and south forks of the Soloman River.

When we left Buffalo Station, we left civilization behind us. We made our own wagon trails, two of which especially were afterwards used by the settlers until the section lines were constructed. One of them ran directly south, crossing Hackberry Creek about fifteen miles from the railroad, at a point where there was a spring of pure water—a rare and valuable find in that region. We camped here many times, and made such a good trail that it was used for years. Our second trail extended across the country, striking Hackberry Creek where Gove City now stands, and led over Plum Creek Divide, whose high ledges of yellow chalk served us as a landmark for twenty miles. From this point we could see Monument Rocks, and near them the remains of an old one-company post on the Santa Fé Trail. Our trail then led up the Smoky Hill to the mouth of Beaver Creek, on the eastern edge of Logan County, and followed the old road as far west as Wallace.

Prairie-dog villages extended west along all the water courses, and open prairies to the state line, and we were rarely out of sight of herds of antelope and wild horses. Near the present site of Gove City, on the south side of Hackberry Creek, there is a long ravine with perpendicular banks ten feet or more in height. This ravine was at that time used as a natural corral by some men who made a business of capturing these wild ponies by following them night and day, keeping them away from their watering places, and giving them no chance to graze, until they were exhausted. They were then easily driven into the ravine and roped; after which they were picketed on the prairie and soon became tame. These wild horses were swift travelers, and the most graceful of all the wild animals of the West, being distinguished for the beauty of their flowing manes and tails.

There was constant danger from Indians, and in order that we might escape as much as possible the eagle eye of some scout who might be passing through the country, our tent and wagon-sheet were of brown duck. This blended with the dry, brown buffalo grass, as we traveled from canyon to canyon, and could not be distinguished very far even by the trained eye of an Indian.

I never carried my rifle with me. I left it in camp or in the wagon, for I soon decided that I could not hunt Indians and fossils at the same time, and I was there for fossils.

I had no unpleasant experiences with Indians, however, although I came very near it once. It was one day late in June, when we were about three miles north of Monument Rocks. A gentle rain early in the morning had taken the glare from the chalk cliffs, and as this is a circumstance favorable to the discovery of fossils, I shouldered my pick and started down the canyon, eagerly scanning the rocks on either side.

About a mile below camp I was startled to come upon a pony trail, so deeply cut into the soft chalk that I knew each horse must be carrying a burden. It had been made within the hour, and as I was anxious to find out what it meant, I took the back trail to the river. There I found that a large band of warriors had sought shelter from the rain in a willow thicket, tying bunches of the twigs together and throwing deer or antelope skins over them to shed the water. They had squatted within these shelters until the storm had passed, and then cooked their breakfasts, as the live coals in many of the ash heaps testified.

There were no squaws or children along; it makes no difference whether women are white or red, they always lose some of their belongings wherever they go, and there was none of such property at this camp. The ponies had been tied to the bushes and not allowed to graze, showing that the party had not expected to camp here, but had simply taken shelter from the rain to avoid the discomfort of traveling with wet buckskin moccasins and leggings. I learned later that it was a large band of Kiowas, Cheyennes, and Arapahoes, under their famous chief, Crazy Horse, going north to join commands with Sitting Bull, in Montana.

The chalk beds which were the field of my labors once composed the floor of the old Cretaceous ocean, and consist almost entirely of the remains of microscopic organisms, which must have fairly swarmed in the water. They were discovered by the late Dr. Bunn, of Lawrence, while a student in the laboratories of the Kansas State University, after Dana and others had said there was no chalk in America.

When the animals that inhabited this ocean died or were killed, their carcasses, buoyed up by the gases that formed after death, floated about on the surface of the water, losing a limb here, a head there, a trunk or tail somewhere else. These detached fragments, sinking to the bottom, were covered by the soft ooze of the ocean floor, and remained there as fossils, while the sedimentary rock was being lifted three thousand feet above sea level.

My explorations began on Hackberry Creek, where I went over every inch of the exposed chalk, from the creek's mouth to its head, in Logan County. Then I searched the river and the ravines that cut into its drainage area along the flanks of the divides.

Perhaps a description of a typical day's experience in one of the long ravines that gash the southern slope of the country may be of interest to my readers.

Human beings, in order to accomplish any result of moment, must be reasonably comfortable, that is, they must not be overhungry or thirsty or sleepy.

If they are, their minds will dwell upon their discomforts, and they will accomplish little, as the hungry boy, who keeps turning his head in the direction of the sun and wondering whether it is not almost dinner-time, is not likely to hoe much corn. My first step, therefore, must be to find water and pitch a camp.

But often I have no idea where water is to be found, and must give as much care to the search as if I were looking for fossils. So while the driver follows me with the wagon, I hunt for water and fossils at the same time.

Both sides of my ravine are bordered with cream-colored, or yellow, chalk, with blue below. Sometimes for hundreds of feet the rock is entirely denuded and cut into lateral ravines, ridges, and mounds, or beautifully sculptured into tower and obelisk. Sometimes it takes on the semblance of a ruined city, with walls of tottering masonry, and only a near approach can convince the eye that this is only another example of that mimicry in which nature so frequently indulges.

The chalk beds are entirely bare of vegetation, with the exception of a desert shrub that "finds a foothold in the rifted rock" and sends its roots down every crevice. This shrub is one of the fossil hunter's worst enemies. Sending its roots down the clefts in the rock, it searches out the fossil bones that have been preserved there, and feasts upon them until they have been entirely consumed, thus thriving at the expense of God's buried dead. More fine fossil vertebrates have been destroyed by this plant than by the denudation of the rock, or the vandal hand of man, although both of the latter have been powerful factors in the destruction of fossils. In those days, however, there were no curiosity hunters to dig up the precious relics, so that they were more abundant than they are now.

All this time I am wandering along the canyon in search of water. Sometimes I come upon gorges only two feet wide and fifty feet deep; sometimes for five miles or more the sides of the ravine will be only a few feet high.

I know that there is water at the river, but it is so far away from my work that I go on and on in the hope of finding some nearer at hand. Dinner-time comes, and the day is so hot that perspiration flows from every pore. A howling south wind rises and fills our eyes with clouds of pure lime dust, inflaming them almost beyond human endurance. Still no water. The driver, with horses famishing for it, makes frantic gestures to me to hurry. To ease my parched lips and swelling tongue, I roll a pebble around in my mouth, or, if the season is propitious, allay my thirst with the acid juice of a red berry that grows in the ravines.

After hours of search, I find in moist ground the borings of crawfishes; with line and sinker I measure the depth to water a couple of feet below in these miniature wells. The welcome signal is given to Will, the driver, and he digs a well, so that both man and beast may be supplied.

If I could sum up all the sufferings I endured in the chalk fossil fields, I should say that I suffered more from the lack of good drinking water than from all the other ills combined. Except when we were in the vicinity of one of the half-dozen springs that are scattered about over an expanse of country a hundred miles long and forty wide, the only water that we had to drink was alkali water, which has the same effect upon the body as a solution of Epsom salts, constantly weakening the system. Yet whole neighborhoods of settlers to this day have no other water for themselves or their beasts, and they show the deteriorating effects in their faces and their walk.

If I have found, scattered along a wash, the bones of some fossil fish or reptile, as soon as we have pitched camp and eaten our meal of antelope meat, hot biscuits, and coffee, we both return with pick and shovel, and, carefully saving each weathered fragment, trace the remains to where the rest of the bones lie *in situ,* as the scientists say,—that is, in their original position in their rocky sepulcher.

Then comes the work in the hot sun, whose rays are reflected with added fervor from the glaring surface of the chalk. Every blow of the pick loosens a cloud of chalk dust, which is carried by the wind into our eyes. But we labor on with unfailing enthusiasm until we have laid bare a floor space upon which I can stretch myself out at full length. Lying there on the blistering chalk in the burning sun, and working carefully and patiently with brush and awl, I uncover enough of the bones so that I can tell what I have found, and so that when I cut out the rock which holds them I shall not cut into the bones themselves.

After they have been traced, if they lie in good, hard rock, a ditch is cut around them, and by repeated blows of the pick, the slab which contains them is loosened.

This is then securely wrapped and strengthened with plaster or with burlap bandages that have been dipped in plaster of the consistency of cream. In the case of large specimens, boards are put lengthwise to assist in strengthening the material, so that it will bear transportation. Later I hope to tell of a method, originated by me, by which the most delicate fossil, even if preserved in very loose, friable rock, may be detached and transported safely.

So, as a hunter will follow the deer, through thickets and over rocks, forgetting hunger and cold and thirst in his anxiety to get a glimpse of his game, that he may add its antlers to his list of trophies, we fossil hunters, Professor Mudge's party and my own, sought our prey over miles and miles of barren chalk beds, cheerfully enduring countless discomforts.

Urged on by enthusiasm and the desire to secure finer and finer material, I went over every inch of the acres of exposed chalk along these ravines and creeks, hoping each moment to find stretched before my delighted eyes a complete skeleton of one of those old sea serpents described by Cope, or a specimen of that wonderful *Pteranodon,* or toothless flying reptile, whose wing expanse was twenty feet or more.

All day, from the first streak of light until the last level ray forced me to leave the work, I toiled on, forgetting the heat and the miserable thirst and the alkali water, forgetting everything but the one great object of my life—to secure from the crumbling strata of this old ocean bed the fossil remains of the fauna of Cretaceous Times.

The incessant labor, however, had a weakening effect upon my system so that I fell a victim to malaria, and when a violent attack of shaking ague came on, I felt as if fate were indeed against me.

I remember how, one day, when I was in the midst of a shaking fit, I found a beautiful specimen of a Kansas mosasaur. *Clidastes tortor* Cope named it, because an additional set of articulations in the backbone enabled it to coil. Its head lay in the center, with the column around it, and the four paddles stretched out on either side. It was covered by only a few inches of disintegrated chalk.

Forgetting my sickness, I shouted to the surrounding wilderness, "Thank God! Thank God!" And I did well to thank the Creator, as I slowly brushed away the powdered chalk and revealed the beauties of this reptile of the Age of Reptiles. Its snake-like tail and flexible movements caused it to appear to Cope a veritable serpent, so that he put it in his new sub-order *Pythonomorpha.*

I well remember the terrible journey over the rough sod to the station with this specimen. I was seized with another attack of ague, and as I jolted about in the bottom of the wagon, I thought that my head would surely burst. Little I cared, though, so that I got my beloved fossil to the Professor.

And I felt amply repaid for my sufferings when the next winter I laid out the skeleton on the platform of St. George's Hall, in Philadelphia, where the Professor spoke for an hour to a spellbound audience, unfolding to them the wonders of the creatures that lived when this old world was young. At the close, which came suddenly, as was usually

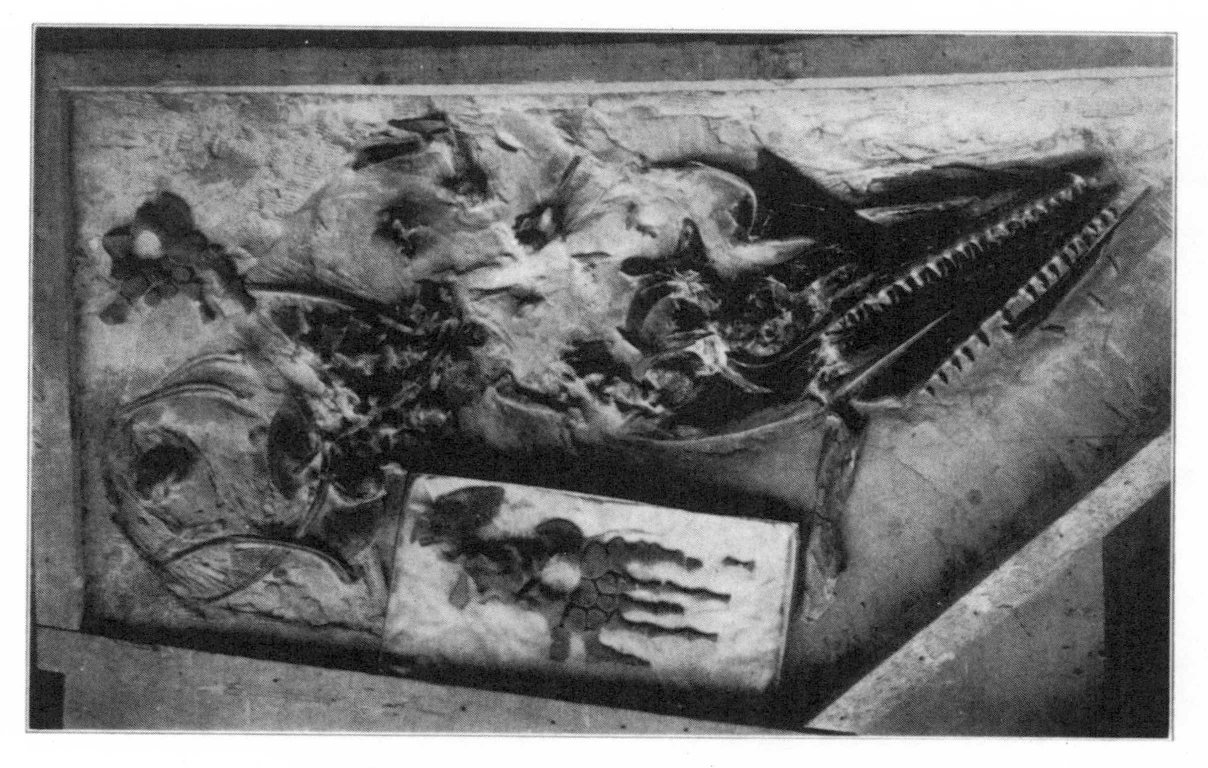

FIG. 6.—SKULL AND FRONT LIMB OF *Clidastes tortor*.
As collected and preserved by Charles Sternberg. (Now mounted in the Carnegie Museum.)

FIG. 7.—SKELETON OF *Clidastes tortor*.
(In American Museum of Natural History.)

FIG. 8.—SKELETON OF RAM-NOSED TYLOSAUR, *Tylosaurus dyspelor*.
(In the American Museum of Natural History.)

the case in Cope's speeches, before the people had had time to come back from the misty past, he turned to where I was sitting on a step, and beckoned me to him. When I got within reach, he turned me around to the audience and said: "Ladies and gentlemen, allow me to introduce to you Mr. Sternberg, the man who found this beautiful example of the fauna of the Cretaceous."

He was much pleased with the hearty applause that greeted me.

This incident illustrates one of the characteristics of Cope which endeared him to all his collectors. He did not think that the money he paid them paid for the dangers and privations they endured, far from their friends and the comforts of civilization. On the contrary, he gave them credit in all his publications for their discoveries of species new to science. And this is the one essential thing to the collector—at least the true collector who values his labor as something that cannot be measured by money. All work done for science has a value above that of money. Lesquereux might have made money if he had remained a watchmaker, and Cope would have won a fortune as a ship-owner if he had entered his father's office, but both men realized that there is work which offers higher rewards than riches; they gave their lives to science, and they will never be forgotten.

But we are far a-field; let us return to the plains and canyons of the Kansas chalk beds.

I recall many trying experiences during that memorable first season. Often we got into barren ground and walked over miles and miles of blistering chalk with nothing to show for our trouble. In one locality the remains might be very abundant, while in another, perhaps just as promising in appearance, thousands of acres would be entirely barren. But we had to go over it all before we could be sure that there was nothing to repay our toil.

Once after two weeks of fruitless effort, we drove into a deep canyon, cut into the upper or reddish chalks near Monument Rocks, which are so much richer in fossils than the yellow or whitish beds farther east.

I had barely pitched the tent and got among the beds when I discovered not only that I was the first collector to visit the canyon but that it was rich in fossil remains. I found two specimens of *Platecarpus,* a species of Kansas mosasaur, in a low knoll, separated by but three feet of chalk.

At the same time one of those uncomfortable cold rains set in, and I was not much encouraged when Will told me that we had no food left. There was plenty of corn for the ponies, but no coffee, flour, bacon, or canned goods, not even an antelope; and we were forty miles from our base of supplies. I would not leave, however, without my load of fossils, as I feared that during my absence my rivals would come upon this Eldorado and clean it out. So the cook was told to parch a kettleful of corn, and we made our meals on that. In fact, we filled our pockets with it and lived on it for three days, eating most of the time to keep ourselves sufficiently nourished.

We had always depended for fuel upon the buffalo chips which even then were strewn about everywhere, but fortunately we found here an old dead cottonwood tree, a rare thing in that region, where even the willows on the river banks are short and stunted. But for this wood we should have suffered.

We remained there until we had loaded our wagon with eight hundred pounds of fossil vertebrates.

During the summer my constant use of a large butcher knife in cutting away the chalk from specimens caused a felon to form in the palm of my hand. A fistula resulted, and for ten days I slept but little, and could not work in the field.

Finally, worn out by hard labor and constant attacks of ague, I felt that my strength was failing, and called on Professor Cope for an assistant. He sent me J. C. Isaac, from Illges Ranch, Wyoming; but matters were not much improved, for Mr. Isaac had but a short time before seen five of his companions shot down and scalped by a band of marauding Indians, and only the swiftness of his horse had saved him from the same fate. Consequently, he saw an Indian behind every bush; and, although I had never been afraid before even when I learned that a large party on the warpath had passed close to my camp, now, worn and tired as I was, I became infected with his fears.

When I found that I could do nothing to get myself out of this mental condition and be of further use to the Professor, I wrote to him, and was ordered home for rest, to meet him later in Omaha, in company with Mr. Isaac.

But before we return to civilization, will my readers go with me on another expedition to these Kansas chalk beds? "How fleet is a glance of the mind!" Instead of an arid, treeless plain, covered with short grass, a great semi-tropical ocean lies at our feet. Everywhere along the shores and estuaries are great forests of magnolia, birch, sassafras, and fig, while a vast expanse of blue water stretches southward.

"But," you ask, "what is that animal at full length upon the water in that sheltered cove?"

Watch it a moment! It raises a long conical head, four feet in length and set firmly upon a neck of seven strongly spined vertebræ. This powerful head terminates in a long, bony rostrum, also conical in shape. Back of the neck are twenty-three large dorsal vertebræ, followed by six pygals, as Dr. Williston calls them, to which the hind arches and paddles are attached. The body terminates in an eel-like tail of over eighty elements, each strengthened by a dorsal spine above and a V-shaped bone, called a chevron, below; so that a vertical section of the lizard would have a diamond shape.

But see! an enemy in the distance is attracting our reptile's attention. It sets its four powerful paddles in motion, and unrolling its forked tongue from beneath its windpipe, throws it forward with a threatening hiss, the only note of defiance it can raise. The flexible body and long eel-like tail set up their serpentine motion, and the vast mass of animal life, over thirty feet in length, rushes forward with ever-increasing speed through water that foams away on either side and gurgles in a long wake behind.

The great creature strikes its opponent with the impact of a racing yacht and piercing heart and lungs with its powerful ram, leaves a bleeding wreck upon the water. Then raising its head and fore paddles into the air, it bids defiance to the whole brute creation, of which it is monarch.

A noble specimen of this great ram-nosed Tylosaur is now mounted as a panel on the wall of the American Museum, in New York, at the head of the stairs on the right (Fig. 8); and a little further on, is a splendid skull of the same species, which I discovered on Butte Creek, in Logan County. Fig. 9 shows a restoration of this species.

Doubtless many of the ankylosed bones which we fossil hunters often find in the chalk of the Niobrara Group of the Cretaceous were broken by blows from these ram-nosed lizards.

We have in Kansas three genera of these mosasaurs as the celebrated Frenchman, Cuvier, named them in 1808. The word literally means a reptile of the Meuse, and it was given them because the first specimen ever found was taken from the quarries under the city of Maestricht, on the River Meuse. For this information, and for much more as to the anatomy of the Kansas mosasaurs, I am indebted to Dr. Williston's splendid work in Volume IV of the University Geological Survey of Kansas: Paleontology, Part I; although, of course, I obtained most of my knowledge from the hundreds of specimens which I collected myself.

Among these are four especially fine specimens, nearly complete, of the flat-wristed *Platecarpus coryphæus* Cope. One of them I sent to the Iowa State University, with head, column, and limbs nearly in position, and still bedded in their native chalk. This fellow, who was over eighteen feet long, must have sunk so deep in the slimy mud of

FIG. 9.—RAM-NOSED TYLOSAUR, *Tylosaurus dyspelor*.
Restoration by Osborn and Knight. (From painting in American Museum of Natural History.)

FIG. 10.—SKULL OF THE FLAT-WRISTED MOSASAUR, *Platecarpus coryphæus.*
(In the Kansas State University.)

the ocean-bed that even the gases formed in his stomach could not lift his body to the surface. A second specimen was sent to the British Museum of Natural History, in London; a third to Munich, Bavaria, and a fourth to the Roemer Museum, in Hildesheim, Germany.

This last specimen is the best I ever took from the Kansas chalk until 1907. It is twenty-five feet long. Unfortunately, the head was all washed away, with the exception of the mandibles and a few bones of the skull. The most remarkable feature of this specimen was the presence, for the first time in my experience, of the complete cartilaginous breast-bone with the cartilaginous ribs, which are very rare. They were described for the first time from the noble Bourne specimen, by Dr. H. F. Osborn, of the American Museum.

This mosasaur, *Platecarpus,* is the most common species known, and is almost as large as the big *Tylosaurus.* It differs from the latter, however, in the shape of the short, strong paddles and the blunt rostrum. The skull in the illustration (Fig. 10) is that of a very fine specimen, one of my discoveries, which was mounted by Mr. Bunker, of the natural history department in the Kansas State University. I have never seen a more complete skull, or one that shows the height so well, in any specimen, unless it is the little *Clidastes velox,* in the Kansas University collection. You will notice the triangular shape of the head, with the strong bones arching back to support the lower jaw by the pulley-like quadrate bone. Notice also that the suspensorium, instead of curving down so that its groove fits over the rounded edge of the quadrate, is straightened out. This is caused by its having been flattened and distorted, as nearly all fossils are, by the immense pressure to which it has been subjected Observe the conical shape of the head in front of the eye-rim, terminating in the hard, blunt rostrum. It is believed by the authorities that a blow from this ram, delivered at full speed, would put an adversary out of commission.

But how did this creature feed itself, when all its teeth are for grasping, none for masticating? And how did it hold its prey, when it has no claw-armed fingers, only weak paddles for swimming?

In answering these questions, we shall describe two characteristics of the mosasaurs which differentiate them from all other reptiles.

If you will look closely at the photograph, you will notice, within the head, and below the eye-socket, a row of recurved teeth. These are the teeth on the pterygoid bones, which are located on either side of the roof of the mouth, near the gullet, and are provided with twelve teeth, more or less. The lower jaw with its powerful sweep on its fulcrum, pressed the living prey firmly upon these teeth so that it could not come forward and escape. Then notice the ball-and-socket joint just back of the tooth-bearing bone or dentary, of the lower jaw. After the wriggling, struggling prey had been fastened on the teeth in the roof of the mouth, the mandibles were shortened by a spreading of this central joint, and the victim was forcibly pushed down the throat.

The species *Clidastes velox* of these Kansas mosasaurs, was, as its name indicates, very agile, with beautiful bones of so firm a texture that they have suffered less than any of the other fossil vertebrates from the vast pressure to which they have been subjected, not only from the enormous amount of material that has been heaped above them, but from the still more powerful upward push which has raised their burial-place three thousand feet above sea level.

I sent one very beautiful specimen of *Clidastes* to Vassar College; so complete, in fact, that it can be made into a panel mount.

I think no artist has more fully appreciated what these great reptiles must have been when alive than Mr. Sidney Prentice, now of the Carnegie Museum, whose beautiful restoration, made to illustrate Dr. Williston's work on Kansas Mosasaurs, is here reproduced (Fig. 11*b*). I am under obligations to him for the labor of his pencil. He has certainly put life into this denizen of the old Cretaceous ocean, and I do not believe that anyone, after a careful study of the skeleton, could find any fault with the restoration, from a scientific standpoint.

In this connection, I should like also to call attention to the beautifully preserved skull I sent to the Carnegie Museum. This specimen shows a complete side view of the head, with mandibles and maxilla, the teeth interlacing as perfectly as in life. The sclerotic plates that protect the eyeball are also in natural position.

The luxuriant life of the Cretaceous ocean was certainly remarkable. Fish swarmed everywhere, and often, as the specimens are uncovered, the scales are picked up by the wind, crumbled into dust, and scattered in every direction.

Among the most common of the fossil bones in those early days were those of a huge fish, whose vertebræ, with fragments of heads and jaws, were found in great abundance, although no perfect specimen has been discovered. Professor Cope, who described this fish, called it *Portheus molossus*. I secured a fine specimen on Robinson's ranch, in Logan County. It lay in a small exposure of chalk along a grassy hill slope, within a stone's throw of the ranch buildings. My son George was my assistant then, and we got out this specimen in the month of

November. Our boarding place was five miles away, and every night the ground froze hard. Nothing daunted, we went to work with a will.

The head and trunk region had already been uncovered, and many of the ribs and spines had been swept away and lost. We took up the head and front fins in a great slab of plaster, as the chalk in which they lay had disintegrated under the influence of the frost. A violent windstorm was raging at the time, and to complete the slab, George had to bring water from a tank a hundred yards away. I can still see that boy running up with his pail of water, trying to carry it so that it would not be emptied by the raging, howling wind that was almost tearing his coat from his back, while I stood and shouted, "Hurry up! The plaster's hardening!"

The rest of the column, to the tail, we took up separately, and as the great tail-fins and many of the caudal vertebræ were present with their spines, embedded in solid chalk, we removed five feet of superincumbent rock, cut a trench around the slab containing the bones, and took it up by digging under it.

This made another huge mass to be handled. The section containing the head weighed over six hundred pounds, and this tail section almost as much. The latter froze solid before we could get it up to the tent, where we kept a fire burning to dry out the water from the bones and thus prevent the injurious effects of freezing. I should like just here to express my gratitude to those ranchmen who gave their time and strength to assist me in handling these huge sections.

When they had been packed with excelsior in strong boxes, a wagon was backed up against the level platform which we had made in throwing out the rock and soil that lay over the specimen. The boxes were then set on edge, and, with the help of boards and rollers, loaded into the wagon for shipment to the railroad thirty miles away.

But my troubles with this specimen were not over; on the contrary, they had just begun. When the section containing the head was being raised on to a table in my shop it fell and its weight was so great that the head was badly shattered, as was the plaster that secured the bones in place below.

Then all through the winter, while I was trying to dry out the specimen, so that it could be cleaned and prepared for shipment, the rats, which inhabited the walls of the laboratory in great numbers, kept pulling out the bran and excelsior that had been put around the delicate bones to protect them; thus causing the broken plaster, with the bones of the head, to sink lower and lower, as the packing was carried away from underneath.

Driven to think out some plan of saving the specimen from destruction, I conceived the idea of shov-

a *b* *c*

FIG. 11.—RESTORATION OF KANSAS CRETACEOUS ANIMALS.
(From drawing by S. Prentice, after Williston.)
a, *Uintacrinus socialis*; *b*, *Clidastes velox*; *c*, *Ornithstoma ingens*.

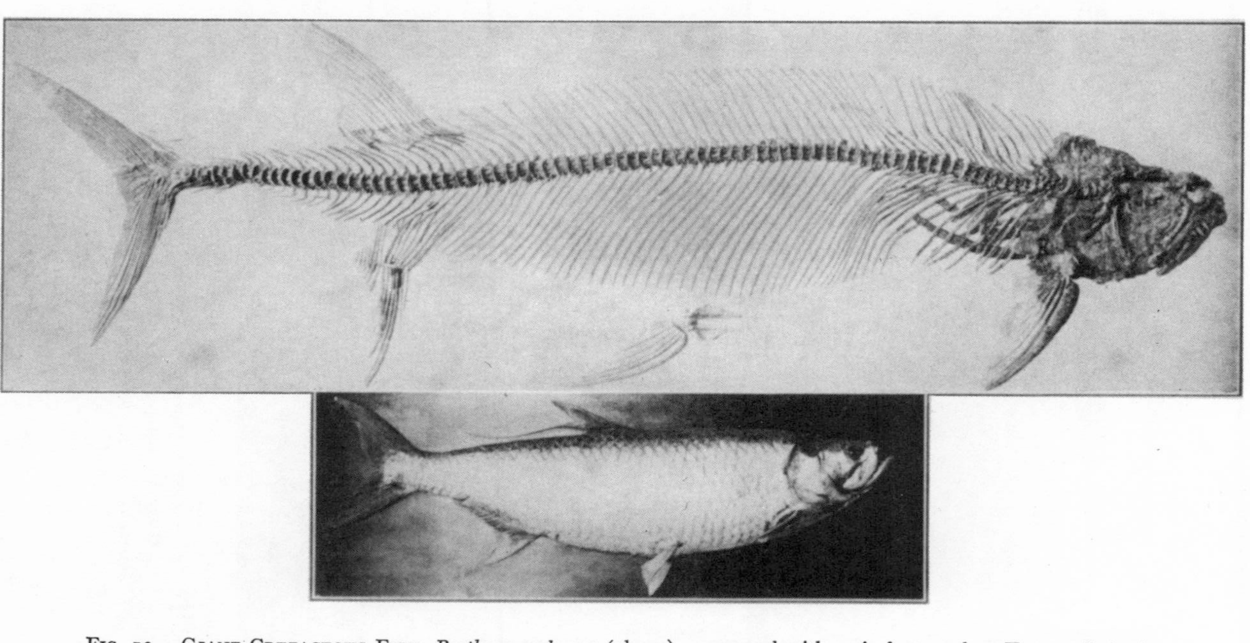

Fig. 12.—Giant Cretaceous Fish, *Portheus molossus* (above), compared with a six-foot modern Tarpon (below). By courtesy of American Museum of Natural History.

ing a number of wooden pegs of various lengths under the broken fragments, so as to push them up into their places and hold them firmly there. All the excelsior was then taken away from beneath them, a frame of lumber made around the section, and the whole space filled with plaster which held all the broken bones in place.

In this specimen I found for the first time a complete column of eighty-five vertebræ, a very important find, as these vertebræ are of so nearly the same size that in restoring an incomplete specimen there was no way of estimating how many of them there ought to be, and for anything to the contrary, one might go on adding them indefinitely, as a certain man in Europe added an enormous number to his mounted specimen of a *Zeuglodon.*

This now famous specimen is mounted above the Bourne Tylosaur, in the corridor of the Halls of Paleontology, at the American Museum. Dr. Henry Fairfield Osborn, in his report describing it, says: "The noble specimen of which a preliminary description is here given, adds another to the many services which Mr. Charles H. Sternberg has rendered to vertebrate paleontology. It was secured by him in the year 1900, near Elkader, Logan County, Kansas. Originally the specimen had been probably complete, but portions of the skeleton, especially the ribs and spines, were injured and partly removed by previous explorers. The fish was purchased by the Museum in 1901, and mounted and partly restored, under the direction of the writer, by Adam Hermann, with the able assistance of Mr. A. E. Anderson. Total length, from tip of tail to a point directly above premaxillaries, 15 feet, 8 inches. Length of skull, 2 feet, 2 inches. Spread of tail, 3 feet, 9 inches." (Fig. 12.)

At the time it was mounted, this great predaceous fish of the Cretaceous was said to be the most striking example of a fossil fish in any museum of the world. Since that day, however, a still finer one has been sent to the Carnegie Museum. This specimen is much superior to that at the American Museum, as the ribs, spines, pelvic fins, arches, and anal fin are in position.

I should certainly be guilty of a great injustice to my friend and the friend of paleontology, Mr. W. O. Bourne, of Scott City, whose name has already appeared in these pages in connection with the great Tylosaur in the American Museum, if I did not give him due credit for his share in the securing of this specimen. He discovered the splendid fish and tumbled a small mountain over on top of it to hide it. Then he kindly gave it to me, and after much digging, my son was able to get trace of it. Mr. Bourne showed his wisdom in thus covering it up, not only from the elements, but also from man, who, out of curiosity, has destroyed some splendid examples of creative power. I shall mention one or two as object lessons before I complete this history.

But let us put life into this fish, whose bones now lie in the Carnegie Museum.

We are back again where the two mosasaurs did battle royal for our enjoyment. Watch that ripple! It is caused by a shoal of mackerel scurrying in toward shallow water, in a mighty column five feet deep. They are flying for their lives, for they have seen behind them their most terrible enemy, a monster fish with a muzzle like a bulldog's, and huge fangs three inches long projecting from its mouth. Two rows of horrid teeth, one above and one below, complete its armature. The great jaws, fourteen inches long and four deep, move on a fulcrum, and when they have dropped to seize a multitude of these little fish, they close with a vise-like power. The crushed and mangled remains pass down a cavernous throat to appease a voracious appetite.

The powerful front fins are armed with an outer ray that moves on a joint in the pectoral arch, a long recurved piece of solid bone, enameled on the outer side and more powerful as a weapon than a cavalryman's sword. This single-edged sword is three feet long, and commands the respect of its owner's enemies, the great saurians, or Kansas mosasaurs. Our fish has only to swim up close to the abdomen of a sleeping reptile, and lay it open for several feet with one sudden stroke. If that is not sufficient, a slap of the powerful tail, with a span of nearly four feet, finishes the work.

But see! nearer and nearer the great fish comes, mouthful after mouthful of the fishes falling into its horrid jaws. It must be starving; so eager is it for its prey that it seems unconscious of the fact that the tide has turned and is moving outward. Now it discovers its danger and turns, but too late. The water has gone back to the deep, leaving it struggling for breath in a shallow pool. It thrashes wildly about with its tail, whose sticky secretions help to envelop it more and more thickly with mud and slime, until at last its struggles cease.

And then the scene changes. The old ocean disappears, and we stand, George and I, three thousand feet above sea level, on Hay Creek, in Logan County, among crumbling ruins of denuded and eroded chalk; and working with pick and shovel in the burning sun, we bring the mighty carcass once more to the light of day.

But I hope to take my readers into this field again, and will pass on now to my expedition in the Bad Lands with Professor Cope.

CHAPTER III

EXPEDITION WITH PROFESSOR COPE TO THE BAD LANDS OF THE UPPER CRETACEOUS, 1876

BOUT the first of August, 1876, Mr. Isaac and I were in Omaha, awaiting the arrival of Professor Cope from Philadelphia.

We met him at the depot, and I remember his watching me with astonishment as I limped along the street on my crippled leg. At last, turning to Isaac, whom he knew to be a horseman, he asked, "Can Mr. Sternberg ride a horse?"

Isaac answered: "I've seen him mount a pony bareback and cut out one of his mares from a herd of wild horses."

That satisfied the Professor, and when we got to Montana, he gave me the worst-tempered pony in the bunch.

We were soon hurrying along over the treeless plains of Nebraska, gaining in altitude every hour, until we reached the highlands of the Great Divide, and plunged down into Weber and Echo canyons, whose forests are dwarfed into miniatures by the majesty of the mountains about them.

It was the first time that I had ever been among these stupendous cliffs and ranges, and I held my breath for very wonder as they unfolded before my astonished vision. They soon became familiar sights enough, but never, even when I gazed every day upon the three Tetons, with the snow glistening in their gorges in midsummer, or upon the mighty ranges of the Rockies, did I lose my feeling of awe at the power here displayed by the almighty Architect who carved these wonderful canyons and set these towering peaks as solemn sentinels over the works of His hands.

We had the pleasure of Mrs. Cope's company as far as Ogden. Then we three men, taking the narrow-gauge railway, went on to Franklin, Idaho. Here the most uncomfortable journey I have ever experienced awaited us,—six hundred miles in a Concord coach, through the dry, barren plains of Idaho. Our six horses raised clouds of fine dust, which penetrated our clothing and filled our eyes and ears, and, sticking to the perspiration that oozed from every pore, soon gave us the appearance of having the jaundice.

I cannot begin to describe the discomforts of that terrible ride. We traveled ten miles an hour, day and night, stopping only for meals, which cost us a dollar each, and consisted of hot soda biscuit, black coffee, bacon, and mustard, without butter, milk, or eggs. If, worn out from continued loss of sleep, we dozed off for a moment, a sudden lurch of the coach into a chuck-hole would break our heads against a post or a neighbor's head. I remember that once when the Professor was almost exhausted from lack of sleep I took his head in my arms and held it there, so that he might get a few hours' rest. I should like here to express my gratitude to the fellow passengers who so often gave me a seat by the driver, where, buttoned in by the leathern apron, I got more than my share of sleep.

When we reached the mountains, the beauty of the scenery and the absence of dust made the journey more endurable, but we had to walk up all the steep ascents.

At Helena we laid off for a few days. There the news was fresh from the battle-field, of Custer and the brave men who had followed him to death. A letter of his, written just before he entered the valley of death, was read to us by the proprietor of the hotel. I remember one sentence of it: "We have found the Indians, and are going in after them. We may not come out alive."

All was excitement, and the Professor was strongly advised against the folly of going into the neutral ground between the Sioux and their hereditary enemies, the Crows. A member of either tribe might kill us, and lay our death to the other tribe.

Cope, however, reasoned that now was our time to go into this region, since every able-bodied Sioux would be with the braves under Sitting Bull, while the squaws and children would be hidden away in some fastness of the mountains. There would be no danger for us, he argued, until the Sioux were driven north by the soldiers who were gathering under Terry and Crook for the final struggle.

Judging from past experience, he concluded that we should have nearly three months in which to make our collections in peace. We would leave the field, he said, when we learned that the great chief was being so closely pressed as to be forced to seek safety in flight to the soil of Great Britain, across the Sweet Grass Mountains into Assiniboia.

His judgment proved good. It was not until November, when a heavy snowstorm had covered both the fossil fields and grass for the ponies, that Sitting Bull gave up the unequal struggle against cold and the Boys in Blue, and retreated to a more friendly soil.

At Fort Benton we found a typical frontier town of that day,—streets paved with playing-cards, and whisky for sale in open saloons and groceries. Our presence had been heralded abroad during our stay in Helena, and the Professor had difficulty in securing an outfit without paying an exorbitant price for it. They knew him to be a stranger, and they "took him in."

Finally, however, he secured four horses for the wagon. The wheelers were worn-out mustangs, which we were obliged to punish constantly to keep at work, while one of the leaders, a fine four-year-old colt, had to be knocked down half a dozen times before he could be taught not to balk and strike out with his fore feet at everyone who came within reach. The other leader, old Major, was as true as steel, and often saved the day, doing his duty nobly in spite of the miserable company in which he was forced to work.

The first night Mr. Isaac and I slept outside the town, with the four wagon horses and the three saddle ponies, which were all picketed with new rope. In the middle of the night, we heard an animal groaning, and rushed out, to find our four-year-old cut fearfully beneath the fetlocks by the ropes. We had to cut him loose, help him up, and bind his wounds. He was able to travel the next day, however, and his accident was not altogether a misfortune, as he was too sore for some time afterwards to show his natural disposition.

We drove down to the mouth of the Judith River, opposite Claggett, where an Indian trader had a store inclosed in a stockade. Here we went into camp. Across the river were the lodges of two thousand Crow Indians, who were preparing for their annual buffalo hunt in this neutral ground, where Sioux and Crow alike buried the hatchet while they hunted the game that was their principal sustenance.

Mr. Isaac, with the dread of the Redman still in his heart, insisted that we must protect the camp by standing guard over it turn and turn about, and to pacify him, the guard was mounted. I took the first turn, and Mr. Isaac the second.

The Professor did me the honor of sharing his tent with me, and we were just dozing off when we heard Mr. Isaac shout "Halt!" Looking out, we saw an Indian approaching, with his squaw behind him, the moonlight bringing out their forms in bold relief.

"Halt! Halt!" called Isaac, leveling his Winchester, but the Indian, followed by his faithful squaw, continued to advance up to the very muzzle of the gun, repeating, "Me good Indian! Me good Indian!"

Cope dressed and went out, and found that the Indian had mistaken us for illicit whisky dealers, and come over to get a supply. The Professor told the man to go to sleep under the wagon, and at daylight to recross and invite half a dozen of the principal chiefs to breakfast with us.

The two Indians lay down and went to sleep as directed, but they had just begun to snore peacefully when Isaac's turn at guard duty was over, and he came to the wagon to wake the cook, a slow, heavy man, whose fat cheeks had induced the Professor to believe that he could cook digestible food. The scout Cope had hired was not on hand, although he, as well as the cook, had demanded his pay in advance before he would accompany us.

After much growling, the cook got up, and remembering that he had left his shoes under the wagon, went to get them and came upon the sleeping beauties. Without more ado, he seized their dirty blanket in both hands and coolly hauled them out on to the open prairie. After which he proceeded to get his shoes.

At four o'clock in the morning it was Cope's turn to go on guard. He was awakened, but as his Spencer carbine was at the bottom of his trunk, and perhaps, too, because he was a Friend, and did not believe in war, he refused to get up; and we slept in safety the rest of the night without a guard.

Just before breakfast the Professor, as was his custom, was washing his set of false teeth in a basin of water, when a party of six stalwart chieftains strode up in single file, in answer to his invitation through the brave we had entertained.

Quickly slipping the teeth into his mouth, Cope advanced with a smiling face to greet his guests, who shouted as one man, "Do it again! Do it again!" He repeated the performance for them again and again, much to their mystification.

After they had tried to pull out their own and each other's teeth, and had failed, they settled down to breakfast. The cook poured out their coffee for them, and when they had had enough they shouted, "When!"

We never knew whether this hospitality was of any benefit to us, as the whole tribe went on their buffalo hunt, and we saw no more of them, but very likely their chiefs forbade petty stealing from our camp, for we lost nothing.

We crossed the Missouri, here a clear, sparkling stream, and the Judith River, and went into camp in the narrow valley of Dog Creek, in the midst of the fossil fields which we had come so far and at such risks to explore.

All about us stretched the interminable labyrinths of the Bad Lands. Above us lay twelve hundred feet of denuded rock, which Cope at that time believed to belong to several formations. The rock consists of great beds of black shale, which disintegrates on the surface into a fine, black dust. The lower levels contain many beds of lignite, which makes a good soft coal, and burns readily. We found beds four feet thick along the canyons. All one had to do was to drive up to the face of the cliff and load a wagon in a few minutes.

As soon as the first streak of daylight appeared, we breakfasted and were off, our picks tied to our saddles, our collecting-bags dangling from the pommels, and a lunch of cold bacon and hardtack in our saddle-bags.

I usually rode beside the Professor, my mount a treacherous black mustang, who was ever on the watch to regain his liberty. A curb bit that almost tore his mouth to pieces was my only means of restraining him. My right ear being totally deaf, I usually rode at the Professor's right, when the trail would admit of our traveling abreast. He was not always in a talkative mood, but when he began to speak of the wonderful animals of this earth, those of long ago and those of to-day, so absorbed did he become in his subject that he talked on as if to himself, looking straight ahead and rarely turning toward me, while I listened entranced.

Not so that wicked black mustang of mine. Suddenly his front feet would leave the ground, and he would stand up at full length on his hind legs. Then feeling the gouging of the Spanish bit, he would drop and run ahead to the Professor's left side. When the Professor, happening to look up, found the place where I had been vacant, he would exclaim in surprise, "Why, I thought you were on my right, and here you are on my left!"

The pony repeated this trick whenever I became so deeply interested in the Professor's talk as to loosen my hold on the reins.

On the very top of the Bad Lands were the Judith River beds, now known, through the researches of the late Professor J. B. Hatcher, to belong to the Fort Pierre Group of the Upper Cretaceous. Here tablelands and level prairies offered plenty of grass for our ponies; so we climbed to these heights, picketed our horses, and went into the gorges in search of fossils. It was necessary to give the loose shale the most careful examination, as only a streak of dust a little different in color from the uniform black around it, indicated where the bones were buried.

As a result of the loose composition of this friable black shale and the overlying rocks of sandstone, the Missouri has lowered its bed twelve hundred feet below the level of the prairies, and the whole country is cut up by a perfect labyrinth of canyons and lateral ravines into a dreary landscape of utter barrenness.

At night the view from above of these intricate passages was appalling. The black material of which the rocks are composed did not permit a single ray of light to penetrate the depths below, and the ebony-like darkness seemed dense enough to cut.

Long ridges, terminating in perpendicular cliffs, whose bases impinge upon the river a thousand feet below, extend back into the country for miles. Often they are cut by lateral ravines into peaks and pinnacles, obelisks and towers, and other fantastic forms. These ridges are so narrow that we could hardly walk along them, and their sides drop at an angle of forty-five degrees. It was only the disintegrated shale on the surface, into which our feet sank at every step, that gave us a foothold and kept us from shooting with frightful velocity into the gorges below.

One day the Professor asked me to climb to a point near the summit of a lofty ridge, crowned by two massive ledges of sandstone, four feet thick, which projected over the steep slope like the window sills of some Titanic building. These ledges, one above the other and separated by sixty feet of shale, had been swept clean for about three feet, so that I found an easy pathway for my feet, when after laborious climbing I reached the lower ledge. From my lofty perch I had a bird's-eye view of mile upon mile of the wonderful Bad Lands, a scene of desolation such as no pen can picture.

It was my duty to search every square inch of the dust-covered slope between the ledges for fossil bones. After much unsuccessful effort, I came to a place at the head of a gorge, where a perpendicular escarpment dropped downward for a thousand feet. The upper ledge of sandstone had broken loose for a space of thirty feet, and this huge mass of rock, four feet thick, carrying with it the loose dirt and polishing the underlying surface as it thundered down the slope, had struck the lower ledge with such force that it too had broken loose and plunged downward into the abyss. A grove of pine trees at the base of the cliff had been crushed to the earth by this avalanche. To my view the remaining trees, which I knew to be about fifty feet high, appeared like seedlings, and the vast mass of rock like a cobblestone.

I concluded that I should have no difficulty in crawling across the smooth space, for I reasoned that if I began to slip, I could drive the sharp end of my pick into the soft rock and thus stop myself. So, climbing up the slope through the loose earth to the base of the upper ledge, I started to cross. When I was halfway over I began to slip, and confidently raising my pick, struck the rock with all my might. God grant that I may never again feel such horror as I felt then, when the pick, upon which I had depended for safety, rebounded as if it had been polished steel, as useless in my hands as a bit of straw. I struck frantically again and yet again, but all the time I was sliding down with ever-increasing rapidity toward the edge of the abyss, safety on either side and certain and awful death below.

I remember that I gave up all hope of escape, and that after the first shock I felt no fear of death; but the few moments of my slide seemed hours, measured by the rapidity with which my mind worked. Everything, it seemed to me, that I had ever done or thought spread itself out before my mind's eye as vividly as the wonderful panorama of the cliffs and canyons upon which I had been gazing a few moments before. All the scenes of my life, from childhood up, were re-enacted here with the same emotions of pleasure or pain. I saw distinctly the people I had known, many of them long forgotten. My mother seemed to stand out more prominently than anyone else, and I wondered what she would think when she heard that I had been dashed to pieces. I even planned how, when I did not return to camp, Cope would set out to find me, following my footsteps into the loose dirt until he reached the slide, and I wondered how he would ever get down into the canyon, and how much of my body would be left for burial.

To this day I do not know how I escaped. I suddenly found myself lying on the ledge, on the side I had left a moment before. Probably some part of my clothing, covered with dust as it was, had acted as a brake upon the polished surface. I lay for an hour with trembling knees, too weak to make my way back to camp.

This experience of mine is another instance of the fact that the human brain forgets nothing, and will yield up everything when the right kind of stimulus is applied.

The excitement of our work and the danger with it seemed to make us reckless of life, Professor Cope even more so than the rest of us, although he was at that time United States Paleontologist, and worth a million dollars. I remember one night he was following a buffalo trail to the river, when suddenly his horse stopped and refused to go further. Without dismounting to find out the cause, he plunged his spurs into the animal, and it sprang into the air. Mr. Isaac, who was behind, followed. The next day they were surprised to find that they had crossed a gorge ten feet wide, and that but for the keen sight and the strength of their horses, they would have been dashed to pieces a hundred feet below.

Cope's indefatigability, too, was a constant source of wonder to us. We were in excellent training, after our strenuous outdoor life in the Kansas chalk beds, while he had just been working fourteen hours a day in his study and the lithographer's shop, completing a large Government monograph, writing his own manuscript, and reading his own proof. When we first met him at Omaha, he was so weak that he reeled from side to side as he walked; yet here he climbed the highest cliffs and walked along the most dangerous ledges, working without intermission from daylight until dark.

Every night when we returned to camp, we found that the cook had spent the whole day in cooking. Exhausted and thirsty,—we had no water to drink during the day (all the water in the Bad Lands being like a dense solution of Epsom salts),—we sat down to a supper of cakes and pies and other palatable, but indigestible food. Then, when we went to bed, the Professor would soon have a severe attack of nightmare. Every animal of which we had found traces during the day played with him at night, tossing him into the air, kicking him, trampling upon him.

When I waked him, he would thank me cordially and lie down to another attack. Sometimes he would lose half the night in this exhausting slumber. But the next morning he would lead the party, and be the last to give up at night. I have never known a more wonderful example of the will's power over the body.

His memory and his imagination, too, were extraordinary. He used to talk to me by the hour, arranging the living and dead animals of the earth in systematic order, giving countless scientific names and their definitions. I forgot the names as soon as I heard them, but the loving tribute which he paid to the wonders of creation has had a lasting and helpful effect upon me. If I ever had any feelings of disgust or fear toward any of God's creatures, I lost them upon a knowledge of the animals as revealed to me by this master naturalist, who saw beauty even in lizards and snakes. He believed, and taught me to believe, that it is a crime to destroy life wantonly, any life. Of course the first law of nature is self-preservation; we must, in order to live, kill our enemies and protect our friends; but this superstitious fear which men and, even more, women have of snakes, lizards, and bugs, how cruel it is! Why should they rejoice when some poor little garter-snake, which has gone as a friend into the cellar walls to destroy rats and mice, is dragged out and cut to pieces? My heart bleeds when I think of the brutal way in which people take life, something they can never give back, and with the great Cope, I cry out against this crime, which is exterminating some of our most beautiful and useful friends. No man can say he loves us, when he wantonly destroys our work; no man loves God who wantonly destroys His creatures.

We found no complete specimens of any fossil

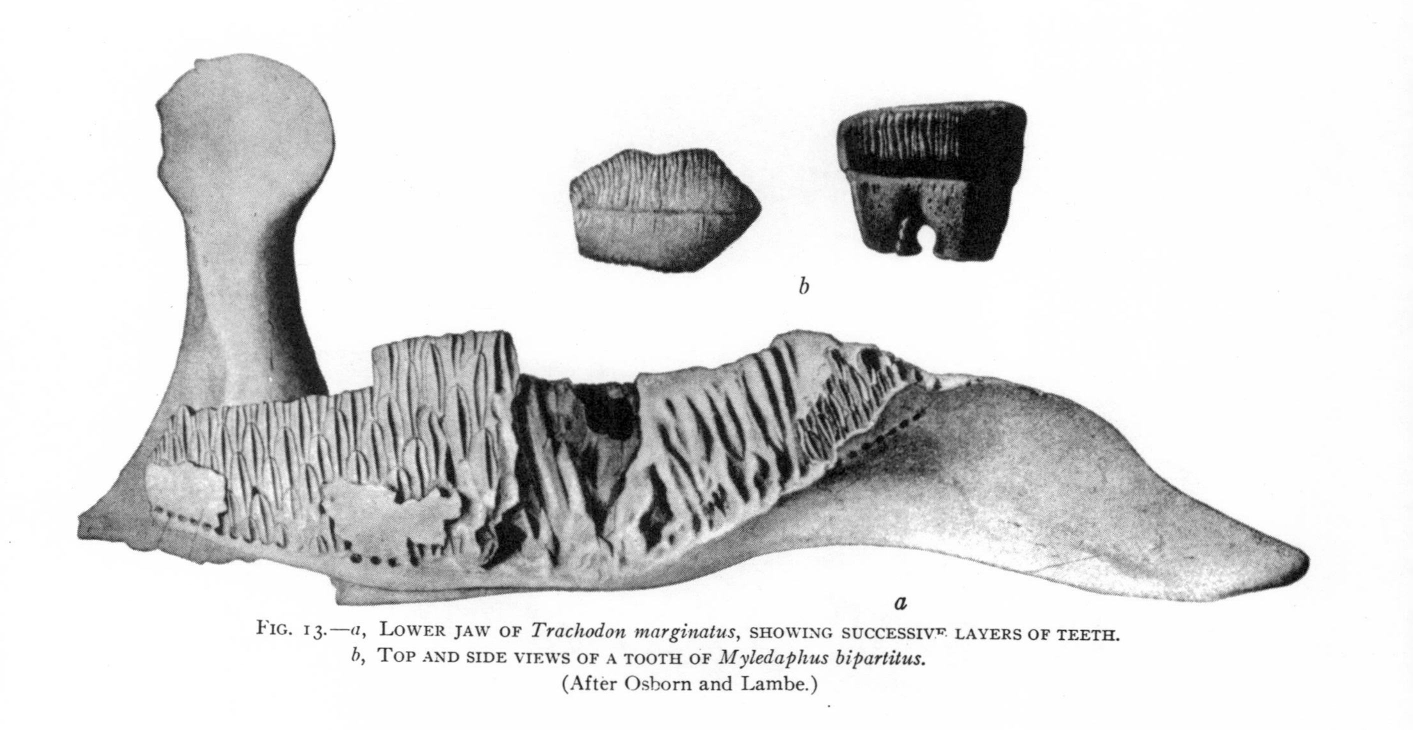

FIG. 13.—*a*, LOWER JAW OF *Trachodon marginatus*, SHOWING SUCCESSIVE LAYERS OF TEETH.
b, TOP AND SIDE VIEWS OF A TOOTH OF *Myledaphus bipartitus*.
(After Osborn and Lambe.)

FIG. 14.—SKULL OF A DUCK-BILLED DINOSAUR, *Diclonius*, FOUR FEET IN LENGTH.
(In American Museum of Natural History.) Photo. by Matthew.

animals during our stay on Dog Creek, but near the summit of the Bad Lands, under beds of yellowish sandstone, we came upon localities literally filled with the scattered bones and teeth of dinosaurs, those terrible lizards whose tread once shook the earth. They are represented now by the little horned toad of central Kansas. Among the fragments were pieces of the finely-sculptured shells of the sea turtles, *Trionyx* and *Adocus*, and remains of that strange dinosaur *Trachodon* (Fig. 13*a*), whose teeth were arranged as in a magazine, one below another, so that when the old teeth wore out, others were ever ready to take their place.

The specimen in the illustration is from Drs. Osborn and Lambe's Contribution to Canadian Paleontology, on the Vertebrata of the Mid-Cretaceous of the Northwest Territory (1902). The splendid Cretaceous dinosaur here illustrated is from Wyoming (Fig. 14). This last form was restored by the late Professor Marsh, and is now mounted in the museum of Yale University. What a strange picture it presents, this great plant-eater, as, standing on its hind limbs, its powerful tail acting as the third leg of a tripod, it grasps the branches of a tree with its weak hands and arms, while its teeth scrape off the tender leaves!

In one of these localities we found teeth belonging to some extinct ray-like fish that were arranged in the roof and floor of the mouth like bricks in a pavement, forming a sort of mill which ground up the shells upon which the creature subsisted. A strange thing about these teeth was that one side of the enamel was white and the other black. Cope called the species *Myledaphus bipartitus* (Fig. 13*b*).

The diamond-shaped enameled scales of the *Lepidotus,* an ancient relative of the gar-pike, were very common, as were also the teeth of several species of dinosaurs besides those already mentioned.

To-day the great museums of the country have complete or nearly complete skeletons of these creatures, the largest land animals that ever inhabited the earth. The splendid specimen of *Brontosaurus* (Fig. 16) in the American Museum at New York is over sixty feet long. Nothing so fires the imagination as a visit to the halls where these ancient lizards now stand.

I am delighted that recent authorities, Drs. Osborn and Lambe, have given Professor Cope credit for these discoveries of his in 1876, discoveries which are made the more memorable by the fact that he was the first scientist who had the foresight and the courage to explore these fossil beds after Dr. Hayden, their original discoverer, was driven out of the region by Blackfeet Indians. Indeed, the chief purpose of this chapter is to put forward the claim that Professor Cope, Mr. Isaac,

Fig. 15.—Professor E. D. Cope.

FIG. 16.—BRONTOSAURUS OR THUNDER LIZARD.
Restoration by Osborn and Knight. (From painting in American Museum of Natural History.)

and myself made the first real collection of these wonderful saurians.

After satisfying himself that there were no skeletons more or less complete on Dog Creek, Cope took the guide and went off down the river to Cow Island, forty miles below. This point was the head of navigation on the Missouri in October, the water then being so low that the steamboat could not get up to Fort Benton. The last boat came up on the fifteenth of October, to carry a load of ore and passengers down to the railroad at Omaha, and as the Professor had decided to take this boat, it was necessary for him to be on hand when it arrived.

A few days later he sent word to us on Dog Creek to break camp and proceed, according to the scout's directions, to Cow Island with all the outfit. This was no easy task; in fact, at first sight it appeared impossible. No wagon had ever before rolled down those steep hillsides. Mr. Isaac, however, took command, and, after removing everything from the wagon except the Professor's trunk, which could neither be packed on a horse nor carried by hand, we began our journey up the long twelve hundred feet to the prairies above.

Working with axes, picks, and shovels, we cut trees, bridged chasms, and made roads, climbing upward step by step, until in the afternoon we reached what for the moment threatened to be the end of our journey. Before us rose the sloping side of a ridge, covered entirely with loose shale, and so steep that it was impossible to climb it even on horseback without making a long diagonal across its flank. At the summit the ridge was narrow enough to be straddled by a wagon, and it sloped down at the same angle on the other side.

The teamster refused to go any further, and this angered Isaac, who said that he would drive himself. So he unhitched the lead horses, and climbing the wagon, urged on the stupid mustangs. One walked in a trail that we had made, the other in the loose dirt below.

I was a good deal concerned as to the fate of both man and team, but experience had taught me the folly of arguing with an angry man; so I sat on my horse and waited for the outcome. Isaac had driven about thirty feet above the level floor, when the inevitable happened. I saw the wagon slowly begin to tip, pulling the ponies over sideways, and then the whole outfit, wagon and horses, began to roll down the slope. Whenever the wheels stuck up in the air, the ponies drew in their feet to their bellies, and at the next turn, stretched out their legs for another roll.

My heart was in my mouth for fear that Isaac would be killed in one of the turns, or that wagon and all would roll over a thousand-foot precipice below, but after three complete turns, they landed, the horses on their feet, the wagon on its wheels, on a level ledge of sandstone, and stood there as if nothing had happened.

When I saw that Isaac was safe, I could not help laughing, and in consequence was told that if I was so smart I could get up the slope myself. I quickly gave orders that the picket ropes be tied together and fastened to the hind axle of the wagon, and that the horses be led singly up the trail. The rope was then carried to the top of the ridge, and the horses were hitched to it, and driven down the steep slope on the opposite side, thus drawing up the wagon. We then righted it so that it straddled the ridge and could be safely hauled out to the level prairie.

After this we had to go back on horses and bring the camp outfit, which we had left at Dog Creek, to the wagon.

About three o'clock that afternoon our scout, who had not showed up during the heavy labor of getting the outfit up to the prairie, was seen coming from the south through a break in the foothills, while at the same time another horseman approached at full speed from the east. At a sign from the scout, our driver stopped his horses, and Isaac and I rested in our saddles.

The second horseman soon proved to be Professor Cope, who galloped up to the guide and stopped him, the gestures of the two men and the sound of their raised voices indicating that an animated argument was going on between them. Finally the scout, his face heated and scowling, came up to the wagon, and without a word, got out his roll of blankets and extra clothing, and started off in the direction of Fort Benton.

The cook shouted after him, and then, springing from the wagon, followed him. When they were out of earshot, the scout stopped, and the two began an excited conversation. Then it was the cook's turn to show of what poor stuff he was made, for, coming back to the wagon, he loaded his blankets and grip on his broad shoulders, and struck out on foot for a wood-camp a few miles to the north, on the river.

When Cope came up he told us that these two men, whom he had paid in full for three months' work, had deserted him here on the open prairie, a hundred and twenty miles from his base of supplies.

It seems that the scout had come across Sitting Bull's war camp, where thousands of warriors, drunk with the blood of Custer and the brave men of the Seventh U. S. Cavalry, were defying the Government in the inaccessible canyons around the Dry Fork of the Missouri. The camp was only a day's journey from us, and the scout and our valiant cook had concluded that their precious scalps were too valuable to risk.

The Professor asked us whether we could carry on the double work which their dishonorable conduct had made necessary, and we willingly undertook to do so, even if it were to mean working our fingers to the bone.

Isaac took the seat, and we prepared to start on, but misfortunes never come singly. Our four-year-old colt, who had had a chance to rest during the delay, suddenly decided that he too would try to put a stop to the expedition. He balked, and when the Professor went up to him to lead him along, he struck out viciously with his fore feet.

Now I imagine that the Professor had put up with about all that he was willing to bear. The cowardly desertion of our men, combined with the discomforts of our situation,—we had had nothing to eat or drink since we left Dog Creek, and the only spring on the route at which we could get good water was miles away,—left little mercy in his heart for this miserable, obstinate horse. He told Isaac to unhitch the animal and tie him to a hind wheel, while I got on top of the wagon, armed with a club to prevent his trying to climb in.

With the whip in one hand, butt end down, Cope approached the horse with the other outstretched, speaking gently to conciliate him. The horse, however, struck out with all his might. Narrowly escaping the blow, the Professor stepped back, raised the whip, and with the butt end, hit the horse behind the ear. The animal fell like a flash, and lay for some time stunned; but when he struggled to his feet, and the Professor approached him again with outstretched hand and soft words, the brute struck again. Again Cope knocked him down, and, although when he rose to his feet, he made another feeble attempt to strike, a third knock-down blow was enough for him. After that he welcomed the Professor's advances, accepting with every symptom of pleasure the caresses bestowed upon him, and when untied, he almost dragged Cope after him in his anxiety to get to his traces. We had no more trouble with him until a long rest and plenty of food caused him to forget his punishment, and made a repetition of it necessary.

It was not until late that night, after fourteen hours of strenuous labor, that we were able to eat our supper of bacon and hardtack, and lie down for a few hours' rest. We slung our food from a tree to get it out of the reach of any grizzlies which might come straying around in search of bread crumbs or bacon rinds. We expected any moment to be rolled out of bed by some prowling paw.

The next day we traveled along through the great level stretches that skirt the Bad Lands. The prairie was covered with thick bunches of grass, and often had been rooted up for acres by grizzlies in search of wild artichokes, a sweet morsel they love. We often saw herds of deer and elk and antelope.

Part of the time our route lay among the foothills of the Judith River Mountains to the south of us; and when we emerged again on to the open plain, we found ourselves in a great amphitheater, a hundred miles across. To the west the towering ranges of the Rockies rose in silent grandeur, their sides scarred deeply with canyons, in whose recesses the white snow gleamed and sparkled in the morning light. To the south, east, and north, the Judith River Mountains, the Little Rockies, Medicine Bow, Bearpaw, and the Sweet Grass Mountains on the border line of Assiniboia made up the circle. A glorious scene! And there was exhilaration too in the thought that ours was the first wagon to roll through these rich solitudes, given up for ages to the red hunter and his game. These hills were soon to re-echo with the shriek of the locomotive, and this rich soil to nourish a thousand souls, but in the days I am recalling, we did not meet a single human being in all the forty miles of our journey.

That night, after another hard day, we halted at the head of a short and very steep ravine ending in an open valley between two ridges, whose lofty precipices abutted on the Missouri twelve hundred feet below.

This valley, Cope told us, was to be our camping ground for some time to come, as a steamboat snubbing-post was situated here. When I learned this, I threw out my roll of blankets and started it on its way to camp. It bounded down the ravine, leaping high in the air from boulder to boulder, and never stopped until it was caught in a bunch of the cactus that covered the level plain below.

Everything but the Professor's trunk was unloaded, and the wagon pulled to the head of the gulch, where Isaac took charge of the tongue, and the Professor and I, each tying a picket rope to the hind axle and making a half-hitch to a convenient sapling, let the wagon slowly down the hill. When the rope was paid out, Isaac blocked the wheels with stones, and we advanced for another hitch, continuing in this way until we reached the bottom. The baggage was then packed down, and, after a space had been cleared of cactus, our tent was pitched. It was not until long after midnight that we sat down to cook our meal, and when we rolled into our blankets we slept the sleep of utter exhaustion.

Not only during this trip, but all through our stay in the Bad Lands, we were tormented by myriads of black gnats, which got under our hat rims and shirt sleeves, and produced sores that gave rise to pus and thick scabs. They got under the saddles and girths too, irritating the horses almost beyond endurance. We were forced, for lack of something better, to cover our faces and arms with bacon grease and to rub the skins of the horses under the collars and saddles with the same disagreeable substance.

Fossil bones always partake of the characteristics of the rock in which they are entombed, and here they were quite hard when we got in to where the rock was compact. The Professor found here the first specimen ever discovered in America of the wonderful horned dinosaurs; *Monoclonius* he called the first species. I assisted him in digging out his specimen of *M. crassus,* a species distinguished by a small horn over each orbit, and a large one on the nasal bones; and I myself discovered two species new to science. One of these, an *M. sphenocerus,* was six or seven feet high at the hips, and, according to Cope, must have been twenty-five feet long, including the tail. It has a long compressed nasal horn, and two small horns over the eyes.

Professor Marsh later discovered a similar form in these same fossil beds, and named it *Ceratops montanus.*

The species I discovered were collected on the north side of the river, three miles below Cow Island, after the Professor had taken the last boat down the river. When we uncovered these bones we found them very brittle, as they had been shattered by the uplift of the strata in which they were buried; and we were obliged to devise some means of holding them in place. The only thing we had in camp that could be made into a paste was rice, which we had brought along for food. We boiled quantities of it until it became thick, then, dipping into it flour bags and pieces of cotton cloth and burlap, we used them to strengthen the bones and hold them together. This was the beginning of a long line of experiments, which culminated in the recently adopted method of taking up large fossils by bandaging them with strips of cloth dipped in plaster of Paris, like the bandages in which a modern surgeon encases a broken limb.

I feel it a great privilege to have been one of the original discoverers of these great horned dinosaurs, whose skeletons are now among the chief glories of our museums.

One day, about the fifteenth of October, Professor Cope, who had been anxiously awaiting the arrival of the last steamboat, concluded to ride out on the open prairie to some bad lands which we had seen on our journey down from Dog Creek. I accompanied him. On the way he fell into one of his frequent absent-minded moods, picturing the land as it must have been at the time of the dinosaurs, when the shale of these black-sided canyons was mud on an ocean floor. So fascinated were we both by his descriptions that the time flew by unheeded, and it was afternoon before we reached the prairie south of Cow Island.

Upon arriving at the bit of bad lands, we separated, agreeing to meet at four o'clock at the place where we left the horses. I kept the appointment, but the Professor was nowhere to be seen, and as hour after hour passed with no sign of him, I began to grow anxious. I knew the foolishness of trying to find him in that network of gorges and ridges, and could only wait, eagerly watching the outlets of the labyrinth.

Just as the sun was sinking behind the Rockies he came out of a narrow ravine with the head of a large mountain sheep on his back. He gave it to me to carry behind my saddle, and with few words we mounted and set off at full speed for home, remembering the three men whom we had met on the prairie at noon, who had been lost for three days in the intricate passages of the Bad Lands. I did not like to think of trying to find the way there after night.

The Professor dashed over the prairie without once drawing rein, clearing bunches of cactus ten feet, sometimes, in diameter, at a single bound; and

I followed suit. So, by a series of leaps, we crossed the ten-mile stretch and drew up at the head of a gorge, from which we could see Cow Island.

Cope eagerly scanned the lights of the little station, and finally decided that a new set had been added to those of the soldiers' tents. He was sure that the long-expected steamer lay at her snubbing-post, and declared emphatically that we must reach Cow Island that night.

I knew the uselessness of trying to combat his iron will, but I pleaded with him against the folly of attempting to thread in the darkness those black and treacherous defiles, where a single misstep meant certain death. I begged him to wait until daylight. We were, to be sure, hungry and thirsty, and food, water, and shelter were to be had only at the river, but sleeping in our saddle blankets without supper was, I urged, preferable to running the risk of being dashed to pieces.

He paid no attention to what I said, but dismounting, led his horse into the canyon. He had to cut a stick to shove in front of him, as his eyes could not penetrate the darkness a single inch ahead. I cut another to punch along his horse, which did not want to follow him.

Sometimes when we had climbed down several hundred feet, the end of the Professor's stick would encounter only air, and a handful of stones thrown ahead would be heard to strike the earth far below. Then we had to turn and climb back through the deep dust to the top, and circling a canyon, plunge down on the other side.

Once we got down to the river four miles from the prairie, and thought that our journey was over, as we could see the lights of the station just across the river. But when we had watered our thirsty horses and started down for the landing, we found our way blocked by a huge ridge with a towering precipice impinging on the river; and we had to drag ourselves back over those four long, hard miles to the prairie, and start again. I freely confess that I should have been willing to lie down in the dust just where I was, and let the horses look out for themselves, but Cope's indomitable will could not be conquered. Back we climbed to the top, and down we went into the next ravine.

I have never known another man who would have attempted this journey. It was both foolhardy and useless, but we could say that we accomplished what no one else ever had in reaching Cow Island through the Bad Lands after dark.

For we did reach it. Just before daylight we got down to the landing across from the station, and sure enough, the steamboat was at her post. But another disappointment was in store for us. The Professor shouted to the sergeant to come and take us over, but his voice was not recognized, and as the sergeant was afraid that the call might come from some Indian who had prepared an ambush, he refused to respond. We were soaked with perspiration, and rapidly becoming chilled by a cold fog that was rising along the shore, and we were obliged to walk back and forth to keep warm until the Professor had recovered his natural voice.

Then, in his haste to correct his error, the sergeant sent a boat across in the wrong place, and it was turned over in the rapids. He had to rescue the half-drowned men, capture the boat, and try again.

At last, however, we were warming ourselves in a tent, where a pot of beans was simmering for the soldiers' breakfast. Not a bean was left when we got through with them, and three pounds of raspberry jam, spread upon, I was going to say a box of, hardtack, followed the beans. Then the sergeant took us both out into the open air and turned back the big black tarpaulin covering the gold ore that was to be shipped to the smelter at Omaha. He made us a warm nest of new blankets, and when we had crawled into it, pulled the tarpaulin back into place. Did we sleep? Ask the deckhands who let the sunlight in upon us about nine o'clock the next morning, when they pulled away the tarpaulin to load the ore.

Cope at once sought the captain of the boat and

said, "I am Professor Cope, of Philadelphia. I have a four-horse wagon at a steamboat snubbing-post three miles below. I would like you to stop there on your way down, and carry my outfit across to this side. My baggage and freight are also there, and I want to take passage for Omaha."

"Well, sir," the man answered, "I am the captain of this boat. If you want to go down the river, you must have your baggage, freight, and self at this landing before ten o'clock to-morrow morning, when I leave for down-river points."

The Professor did not argue the question further. He tried to get the loan of an old sand-scow, but the man who owned it had heard this conversation with the captain, and refused to lend it. The Professor was obliged to purchase it for an enormous price, and the next day left it where he got it. We boarded this scow, and leaving our ponies picketed across the river, paddled down to camp, where, to our disgust, we found that Mr. Isaac had gone out into the Bad Lands to look for us. There was no time to lose; so, although stiff and sore from our night's exertions, we plunged into the work of lowering the tent, packing our stores and fossils into the wagon, and dragging everything aboard the scow. We were ready to start when Mr. Isaac appeared.

We crossed the river, swimming our horses;

and then came the time for old Major to go it alone and show his worth. We converted the Missouri into a canal, and its northern bank into a towpath. Old Major we hitched to a line attached to the scow; and while a couple of mountain men whom we had in camp kept the boat away from the shore with long poles, I rode the big horse, often right into the river, until he began to sink in a mud bank, and I had to turn hastily back to shore. The Professor and Mr. Isaac had the worst places, for they had to keep the rope from being caught by a snag or rock; and when it did catch, if they did not instantly loose their hold upon it, the tension threw them far over into the river, and they had to get out as best they could. This occurred a number of times.

When about sundown we hove-to under the big steamer, the deck was crowded with passengers watching our approach. Cope was covered with mud from head to foot, and his clothing, with hardly a seam whole, hung from him in wet, dirty rags. He had forgotten to bring along any winter wearing apparel, so, although the nights were quite cold, and the women were clad in fur coats and the men in ulsters, he emerged from the sergeant's tent, whither he had carried his grip, in a summer suit and linen duster.

He told me about a funny experience that he had on the boat on the way down the river. It goes without saying that in that long trip he taught the passengers more natural science than they had ever learned in all their lives before. At a certain wood-camp, he and some others went ashore and found the skull of a Crow Indian. The Crow method of burial was to wrap the body in a blanket, lay it on the ground, and build around it an open frame of logs, to keep away wild animals. It was an easy matter to pick up a skull.

The Professor carried his find aboard in his hands before everyone, and was beginning to tell his enlightened listeners the special cranial characteristics of this tribe, when a body of deckhands, headed by their appointed speaker, came forward and told the captain that they would not allow Professor Cope to "emulate the dead." He must take the skull back to its grave or they would not remain aboard and take the boat down to Omaha.

"Why," said the speaker earnestly, "we will be caught on every mud bank in the river, and there is no telling what calamities will happen, if he is allowed to emulate the dead."

There was no getting them to back down from their position, and the Crow's skull was restored to its grave. But the Professor said afterwards, "We had about a dozen skulls packed in with the fossils, and in spite of them, reached Omaha without having to walk on stilts, as had been prophesied."

Shortly after the Professor left us, I discovered a fine specimen, one of those mentioned earlier in this chapter, three miles below Cow Island, near the base of a high tableland, where I kept my pony picketed while I worked. One day, when I prepared to mount him, I noticed that he was unusually quiet. His custom was to start on a run as soon as my foot touched the stirrup, leaving me to get into the saddle as best I could. This time he stood still, and when I reached my seat and lifted the lines, I found that they were perfectly useless, as the curb was broken.

Before I could dismount, the brute started at a rapid pace across the tableland toward a sheer precipice, hundreds of feet high. I settled myself firmly in the saddle and hung on with both hands to the hand-holds behind, fearing that he might try to hurl me over; and that was just what he did. When he got within a few inches of the brink, he planted his feet and stopped suddenly. But Providence and long practice in riding all kinds of horses enabled me to keep my seat, and fortunately, the saddle girths held.

I was just about to dismount, when suddenly the determined animal whirled around and started for the precipice on the other side, where he went through the same performance. And not satisfied even then, tried the trick a third time. Then he allowed me to dismount and mend the curb. In payment for his treachery, I forced him to run at full speed down the steep and rugged trail to camp.

This chapter has been largely taken up with adventures and a study of the man Cope; but as a matter of fact, there was little else to tell about, as we were in such haste that we secured few specimens, and the most important result of the expedition was our discovery of many new specimens of dinosaurs, represented chiefly by teeth.

On the first of November a heavy snowstorm set in, promising to leave the country covered with snow for the winter; so we loaded our outfit and started for Fort Benton. The sergeant went with us, very fortunately, as it proved; for one night, as we were camping in the Bear Paw Mountains, one of our crazy mustang wheelers heard a wolf howl and started on a run for one of the other horses which was picketed farther down the slope. Coming suddenly to the end of its rope, its feet slipped, and it fell and broke its neck. But for the sergeant's horse we could not have hauled in our load.

Countless herds of buffalo were being driven to the Bad Lands by the storm, as were also great droves of deer, elk, and antelope. It seemed as if it would be impossible to exterminate them. Yet I learned by the papers the other day that the last herd of buffalo of any size had been sold at three hundred dollars a head to the Canadian Government, Uncle Sam being too poor to make the purchase.

We reached Fort Benton in safety, learning later that Sitting Bull had crossed at Cow Island and killed the soldiers who had been left there. I never saw my associate, Mr. Isaac, again, but I know that he discovered some fine material the next year.

I made the return stage journey of six hundred miles in six days. Through the mountains the thermometer averaged twenty below zero, and I ate four hearty meals a day. I recrossed the Great Divide on the Union Pacific Railroad, made a brief visit home, and went on to spend the winter with Professor Cope.

CHAPTER IV

FURTHER WORK IN THE KANSAS CHALK, 1877

I SPENT the winter of 1876-77 with Professor Cope, first at Haddonfield, then at his new home on Pine Street, in Philadelphia.

At Haddonfield the commodious loft of a large, old-fashioned barn was fitted up as a workshop, and I had also a bed here. I boarded with a Mr. Geismar, Professor Cope's preparator, but I had a standing invitation to eat dinner every Sunday with the Professor and his wife and daughter, a lovely child of twelve summers.

I shall never forget those Sunday dinners. The food was plain, but daintily cooked, and the Professor's conversation was a feast in itself. He had a wonderful power of putting professional matters from his mind when he left his study, and coming out ready to enter into any kind of merrymaking. He used to sit with sparkling eyes, telling story after story, while we laughed at his sallies until we could laugh no more.

I never knew his wit to fail him. I remember being present at a meeting of the Academy of Science, in Philadelphia, at which he was up for re-election to the office of recording secretary, and was defeated. Among others, Professor William Moore Gabb made some remarks against him. Cope's only defense was "Now, William, more gab!"

I attended also the dinners which he gave to his hosts of friends in the city, and the luncheons at which Mrs. Cope entertained the young men to whom the Professor gave lectures in his own home. He told his funniest anecdotes on these occasions, and used to call on me for my story of the old farmer who, while at work hoeing corn in a stump-field on the side of a hill, saw a hoop-snake at the top take its tail in its mouth and begin to roll down towards him. Springing behind a stump, he struck at it with his hoe handle, into which the sting at the end of the snake's tail entered deeply. In less than an hour the handle had swelled up to the size of a man's leg.

I believe that this story-telling of which he was so fond was for Cope a form of relaxation from his heavy work in the study, and that his ability to give himself up so thoroughly to it in his leisure hours was what enabled him to accomplish in his life an amount of work such as few men have ever accomplished. It would take a volume even to name the titles of all the products of his industrious brain. One of them alone, the great Volume III of the "Tertiary Vertebrata," often called "Cope's Bible," has over a thousand pages of text, beside many fine plates. It was published by the Government, in 1884.

Before starting back to outfit another expedition to the Kansas Chalk, I secured the services of Mr. Russell T. Hill, an able young man who was working in the Academy under the Jesup Fund; and upon our arrival at Manhattan, I hired Mr. A. W. Brouse as teamster and cook.

About the last of March we started with a team of ponies and a light spring wagon upon our long and extremely tedious journey across the state of Kansas, to our headquarters at Buffalo Park. At Chapman Creek, a few miles from Junction City, we were stopped by high water. A raging torrent twenty feet deep filled the bed of the creek; neither man nor beast could have crossed it alive. We were, therefore, horrified to see a farmer, sitting on a seat on top of two sets of side-boards in a lumber wagon, come driving down into this fearful flood. I called to him to stop, and asked him what he was going to do.

"I must come over," he shouted.

"Why," I answered, "the water is twenty feet

deep, and running like a mill race. You'll be swept away."

"But I have not had my mail for a week. I must come over," he shouted back.

"Well," said I, "you big fool, why don't you go down to the railroad bridge, just below here, and walk over?"

"By Chimmeny," he said, "I hadn't thought of that!"

As we were now in the antelope country, we were rarely out of antelope meat. One morning we saw a buck antelope standing close to the railroad track, watching an incoming train. I remarked, as I urged the driver to hurry up his horses, that perhaps someone would shoot the animal from the train. And sure enough, as the train passed, a window flew up, and a man with a revolver shot the buck through the neck. It began to describe a circle, its feet planted together, and springing from the wagon, I cut its throat with a butcher knife, while the boys held its horns.

Another time, as we were traveling along over the prairie, we suddenly came upon a young antelope hidden securely in the center of a bunch of grass. We should not have seen him at all from the ground, but being above him on the wagon seat, we looked right down on him. The boys jumped out, and approaching the little chap carefully, were just spreading out their arms so as to be ready to grab him, when he sprang to his feet so quickly that their hands were thrown into the air, and darted off. The boys started after him at the top of their speed, but they might as well have tried to catch a streak of lightning.

One day we were camping at the spring on Hackberry, south of Buffalo, when a couple of men rode up to us. They said that they were cowmen, and that they had lost their outfit. I invited them into my tent, and after supper gave them the boys' bed, the boys themselves climbing into the covered wagon.

Early in the morning one of the men wakened me and asked for a revolver. There was an antelope in camp, he said. I handed him a Smith and Wesson, and peeped out, to see a fine buck standing just at the end of the wagon tongue, looking over the tent and wagon. The stranger opened fire at three or four paces and emptied the revolver. Then throwing it down as of no account, he asked for a gun. I gave him a Sharp's rifle and a cartridge belt. In the meanwhile the antelope had walked a few yards away and turned to look at us. The man fired several shots, and threw down the rifle also, and as the boys were by this time climbing out of the wagon, one with a Winchester, the other with a little Ballard, he borrowed from them first one firearm and then the other, and blazed away without once drawing blood. Finally the buck deliberately moved over the hill and out of sight, while the man swore that it had a charmed life. We thought otherwise, however, and the boys followed it; soon returning with it swinging from a gun, which they carried on their shoulders like a pole.

I recall another ludicrous incident connected with this expedition. We happened to be at Buffalo Station once when Professor Snow, the much-loved Kansas naturalist, and at one time the chancellor of the State University, was in town with a large party of students, on his annual insect hunt.

The old Chisholm cattle trail led through Buffalo, and one day the owner of a large herd of Texas cattle, who was passing through, noticed Professor Snow and his party out on the prairie with their nets in their hands, running about as if possessed. It happened to be the first time that he had ever seen insect collectors at work, and his curiosity was aroused.

"What are those men doing?" he asked Jim Thompson, the storekeeper.

"Catching bugs," was the laconic reply.

"I don't believe it," said the cowman. "They are grown men."

"All right," said Jim, "you can find out for yourself if you want to."

The man started off after the Professor, and I waited, with a good deal of curiosity, to hear his report of the conversation. On his return he was in a brown study. The Professor had taken him into his tent, and shown him hundreds of mounted insects, reeling off their names to him until his head whirled.

"Well, did I tell you the truth?" Jim asked.

"That man," said the cowman, "is the smartest man I ever saw. He knows the names and surnames of all the bugs in this country."

On the thirtieth of April we drove down to the Smoky, thirty miles south of Buffalo, and got caught in a quicksand, but managed to save both team and wagon. We camped at the mouth of a large ravine with plenty of grass in it.

All that night it blew a perfect gale. Did you, dear reader, ever try to sleep in a tent when the wind was high and the canvas flapped about you, waking the fear that at any moment the pegs might pull out or a seam part? Do you know what it is to lie, deafened by thunder and blinded by lightning, while the rain and sleet dash against the thin covering which is all that separates you from the fury of the storm? It is not a pleasant experience, and yet in all the years that I have gone camping, although I have expected time and again to find my tent torn to shreds over my head, my fears have never once been realized. Even in the most terrible storms my tent has stood securely, and I have escaped without serious inconvenience.

On this trip, however, we did have a disagreeable experience. A cold rain continued for four days, and the tent sprang a leak right over my bed. Moreover, the buffalo chips were so wet that we could not build a fire, and had to eat cold food and sleep in wet blankets.

Among the difficulties with which we had to contend on this expedition was a defective wagon wheel. One day, as we were driving along a slope, our lower wheel dished out, and dumped us, load and all, to the ground. Upon examination, we found that the maker had used a hub whose mortises were too large for the spokes. The latter had been held in place by wedges which had been painted over so that they should not be detected. The man who sold us the wagon had guaranteed it for a year, but unfortunately, he lived two hundred miles away. When the necessity arises, however, one can solve any problem somehow; so we took off the tire, put back the spokes and wedges, heated the tire in a fire of buffalo chips, and reset it. We tried to drive carefully after this and avoid sloping places, but it generally happened that when we least expected it, we would fall by the wayside. Most aggravating of all, when we did take the defective wheel back to the man who guaranteed it, he gave us another even more unreliable than the first. It is a mystery to me how manufacturers can play such miserable tricks on their customers.

We were much inconvenienced also by the illness of one of our horses. He often gave out on the open prairie, in one case, I remember, three miles from water. The only vessel we had in which to bring it to camp was a gallon jug, and it kept one person busy getting enough for our use. We were finally obliged to get another horse in place of the sick one; and our bad luck persisting, hit upon one which had evidently been trained to the wheel of a coach, for as soon as the last trace had been hitched, he was off like a shot. Fortunately, his mate could not run as fast, so that they simply went round in a circle, and the boys, watching their chance, caught hold of the wagon and got aboard.

This horse was continually giving us trouble. One day when we were about to cross Hackberry Creek I went ahead with my pick and struck the dry, cracked clay of the bed, to see whether it would hold. As I could not break through, I concluded that we could cross safely, and beckoned to Will Brouse to come on. Whereupon that miserable mustang, taking his bit between his teeth, came down the hill with the load at full speed, and, dashing onto the hardened clay, broke through into the thick mortar below.

The boys, jumping out, managed to get both horses unhitched before they went down, and quickly hitched them to the hind axle of the wagon, to save the load of fossils which we were hauling to the station. Then began a performance of that tantalizing trick which horses know so well how to play. Rowdy would make a rush forward, as if he intended to haul out the load in a hurry, but the moment he felt the collar press his neck, he would fall back against the wheel, while his mate went through the same performance. So they see-sawed up and down, until I could stand it no longer, as the wagon was slowly sinking. I took the lines, and putting all my will-power into the command "Get out of this!" I forced them to pull together and haul the wagon out to solid ground. Then when we unhitched them, they ran away and scattered singletrees, nuts, and bolts all over the prairie.

South of the river we found some fine examples of large *Haploscapha* shells, some of them a foot in diameter. The valves of this shell are shaped a little like a woman's bonnet, and the name Conrad gave it, "*Haploscapha grandis,*" may be freely translated "The great hood." (Fig. 17.)

We found many fish and saurians or mosasaurs also. Very different was our method of collecting

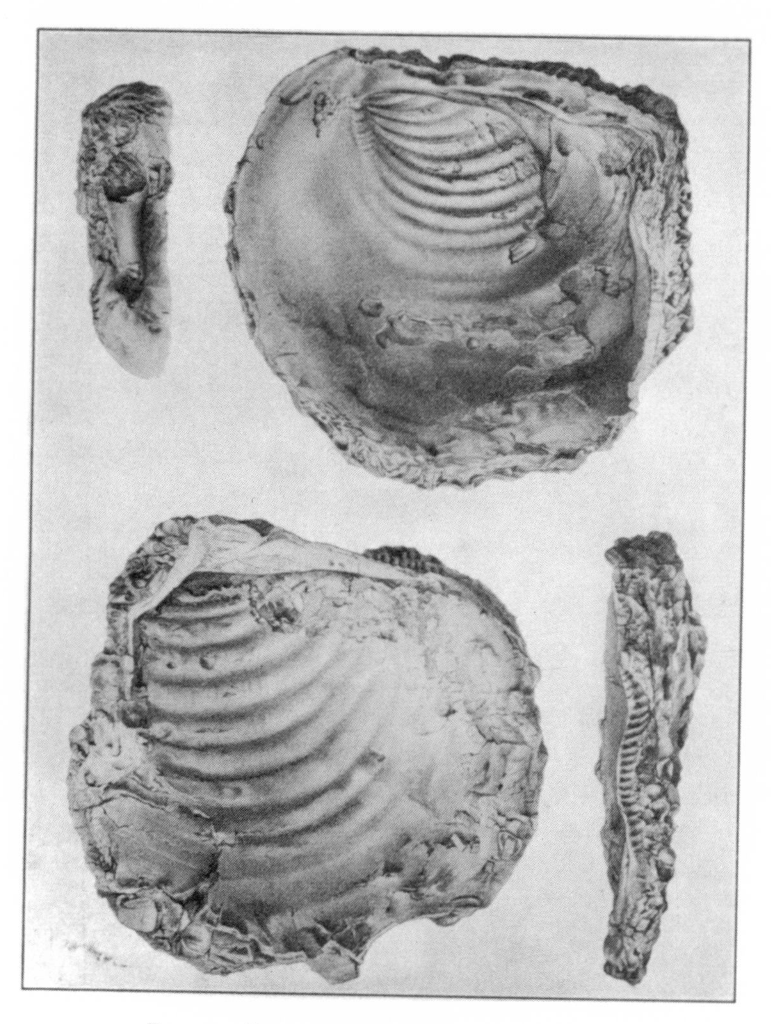

FIG. 17.—FOSSIL SHELLS, *Haploscapha grandis*. (After Cope.)

FIG. 18.—CHARLES STERNBERG AND SON TAKING UP A LARGE SLAB OF FOSSILS FROM A CHALK BED IN GOVE CO., KANSAS.

FIG. 19.—CAMP AND WAGON OF THE FOSSIL HUNTERS ON GRASSWOOD CREEK, CONVERSE CO., WYOMING.

them then from what it is now, for fossil hunting is as capable of improvement as any other form of human endeavor. Then we went over, in a few months, all the chalk in western Kansas, which lines the ravines on either side of the Smoky Hill and its branches for a hundred miles; now it takes us five years to get over the same ground. Then we dug up the bones with a butcher knife or pick, and packed in flour sacks with dry buffalo grass, which we pulled with our fingers. Some strange animals were created by Cope and Marsh in those early days, when they attempted to restore a creature from the few disconnected bones thus carelessly collected. Now we take up great slabs of the chalk, so that we can show the bones *in situ*, that is, in their original matrix, so that they may be the more easily fitted together in their natural relations with each other.

When, after much careful exploration, we find, sticking out of the edge of a canyon or wash, the bones of some "ancient mariner" of the old Cretaceous ocean, we first lay bare a floor above the bones by picking away the rock. Then I, usually stretched at full length on this floor, with a crooked awl and a brush, uncover the bones enough to be able to determine how they lie, often keeping up the tedious work for hours. When the position of each bone has been ascertained, my son George, who for years has been my chief assistant, and I cut trenches around the specimen, and, hewing down the outside rock two or three inches, make a frame of 2 x 4 lumber, cover the bones with oiled paper, and fill the frame with plaster. As the fossil rarely lies level, it is necessary to have the cover ready to nail on, a board at a time, while the plaster is being poured in. This results in a panel of even thickness, with every bone in or near its original position, or at least in the position in which it was buried.

After the plaster has hardened comes the difficult labor of digging the rock away from underneath. One has to lie on one's left side and work with a light pick, using great care, so as to cut away the rock just enough to allow the frame to come down by its own weight. If force is used very likely the rock, with its enclosed fossil, will be torn from the frame, and the specimen ruined. Afterwards the rock is leveled off even with the frame, and the bottom nailed on. The case is then placed in a larger box with excelsior carefully packed around it.

The illustration (Fig. 18) shows a huge panel in process of being cut out. George and I spent two weeks of heavy labor upon another. Luckily, it was preserved in chalk hard enough to allow of its being lifted without breaking. The slab was about four inches thick, and weighed at least six hundred pounds, yet he and I handled it entirely alone, getting it boxed and into the wagon ourselves.

My old friend, Dr. S. W. Williston, who in the seventies was in charge of collecting parties for Professor Marsh, and is now a noted authority in paleontology and professor of that science in the University of Chicago, describes this specimen in his great work on North American plesiosaurs, a Field Columbian Museum publication. He says: "The specimen of *Dolichorhynchops osborni*, herewith described and illustrated [Fig. 20], was discovered by Mr. George Sternberg, in the summer of 1900, and skilfully collected by his father, the veteran collector of fossil vertebrates. The specimen was purchased of Mr. Sternberg in the following spring for the University of Kansas, where it has been mounted and now is. When received at the museum, the skeleton was almost wholly contained in a large slab of soft, yellow chalk, with all its bones disassociated, and more or less entangled together. The left ischium, lying by the side of the maxilla, was protruding from the surface, and part of it was lost. The bones of the tail and some of the smaller podial bones were removed a distance from the rest of the skeleton, and were collected separately by Mr. Sternberg. The head was lying partly upon its left side, and some of the bones of the right side had been macerated away. The maxilla indeed had disappeared.

"The task of removing and mounting the bones

has required the labor of Mr. H. T. Martin the larger part of a year, and is as finally mounted, an example of great labor and skill on his part. . . . The skeleton, as mounted, is just ten feet in length. The neck in life must have been thick and heavy at the base. The trunk was broad; the abdominal region short between the girdles; the short tail was thick at its base. The species was named in honor of Professor H. F. Osborn, of Columbia University."

In his introduction Dr. Williston speaks of the great scientific value of this specimen of the plesiosaurian family, of which he says: "Thirty-two species and fifteen genera have been described from the United States, and in not a single instance has there been even a considerable part of the skeleton made known."

I am glad that the University of Kansas owns this splendid denizen of her ancient Cretaceous sea.

My collection in the Royal Museum of Munich is said by Dr. H. F. Osborn to be the finest prepared collection of Kansas Chalk and Texas Permian vertebrates in the world. A recent letter from my friend Dr. Broili, an assistant there, says that the collection contains over eighty-five distinct species of extinct vertebrates. Among these, there are eighteen species and seven genera new to science. Seven papers have been published describing this material, by J. C. Merriam, A. R. Crook, Charles R. Eastman, F. B. Loomis, F. Broili, L. Neumayer, and L. Strickler, respectively; and it has been illustrated by forty plates. The lamented German paleontologist, Dr. Carl von Zittel, under whom I served the Munich museum for several years, wrote me that I had erected here " an immemorial monument " to my name.

Here rests, far from its native shores, the most complete skeleton of the Cretaceous shark, *Oxyrhina mantelli* Agassiz, ever discovered in any formation. It formed the basis for the inaugural address delivered by Charles R. Eastman before the Ludwig-Maximilian University of Munich.

I discovered this specimen while conducting an expedition for Dr. von Zittel. I was entirely alone, and camping on one of the ravines that score the southern slope of the Smoky Hill valley, south of Buffalo Park. I had already found a number of flattened disks, the centra of fish vertebræ, which Dr. Williston had assured me belonged to a species of shark, as he had found teeth associated with them. I was delighted, therefore, to find here a continuous string of them leading into a low knoll. I quickly shoveled away the loose chalk and cleaned up the floor, to find the whole column, nearly twenty in length; while the skull was represented by great plates of cartilaginous bone, containing some two hundred and fifty teeth from the roof and floor of the mouth. The larger teeth were over an inch long and covered with a shining, dark-colored enamel. They were as sharp and polished as in life, and lay in or near their natural positions.

This is the first time and, I believe, the only time that so complete a specimen of this ancient shark has been discovered. The column and other solid parts were composed of cartilaginous matter which usually decays so easily that it is rarely petrified. I suppose my specimen was old at the time of its death, and bony matter had been deposited in the cartilage. It is not very likely that such a specimen will ever be duplicated. Dr. Eastman's study of this skeleton enabled him to make synonyms of many species which had been named from teeth alone.

Among the most valuable of my further discoveries in the Kansas chalk beds was that of two nearly complete skeletons of that great sea tortoise, *Protostega gigas* Cope. The type had already been described by Professor Cope from a number of disconnected bones which he found near Fort Wallace in 1871.

In 1903 I was so fortunate as to find a practically complete skeleton of *Protostega gigas* in normal condition, that is, with the bones all in or near their original positions. The late Dr. J. B. Hatcher,

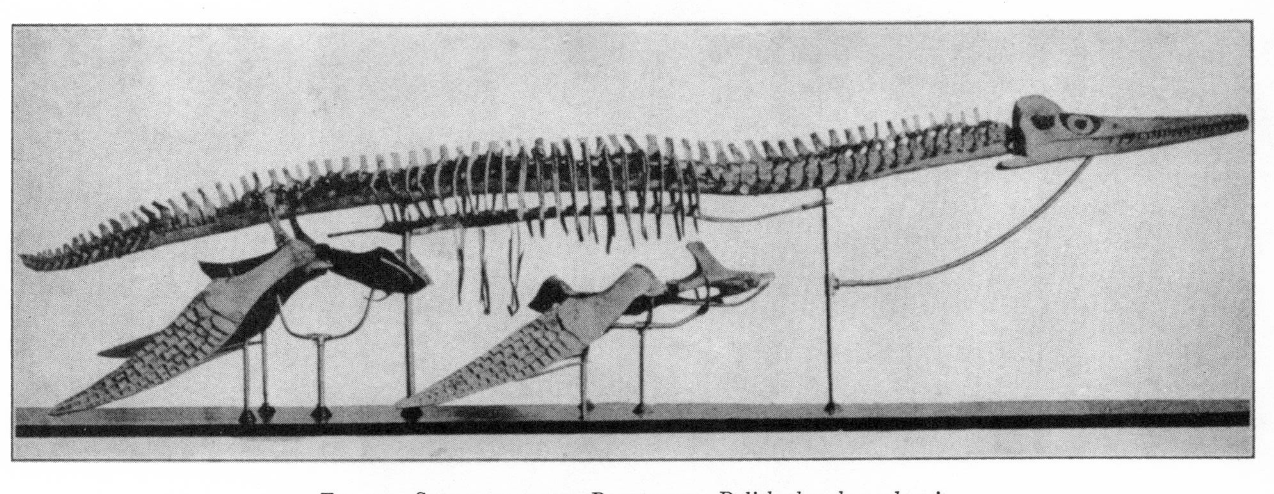

FIG. 20.—SKELETON OF THE PLESIOSAUR, *Dolichorhynchus osborni*.
Discovered by George F. Sternberg and collected by Charles Sternberg. After Williston. (Now in the Kansas State University.)

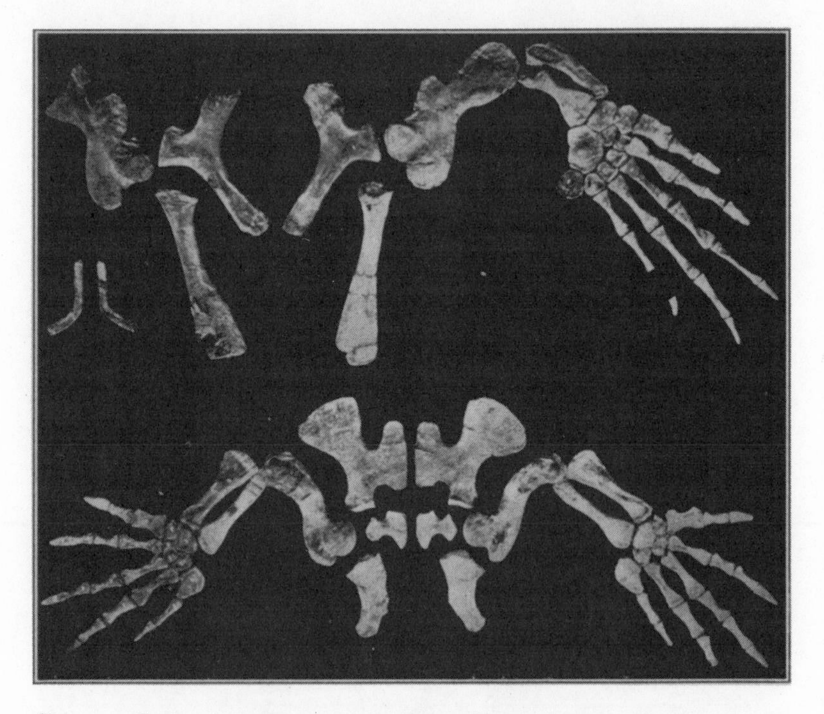

FIG. 21.—FOSSIL LIMB BONES OF THE GIANT SEA TORTOISE, *Protostega gigas.* Collected by Charles Sternberg.

whose death in the very noonday of his glorious career as a fossil hunter cast a gloom over the world of paleontology, purchased this specimen from me for the Carnegie Museum. It has been described in the Memoirs of the Carnegie Museum by Dr. G. R. Wieland, the authority on extinct turtles, under the title " The Osteology of *Protostega*." He says, on page 289: " The third of a century which elapsed since Cope's discovery of *Protostega gigas*, has not sufficed to bring forth a complete restoration of any single individual of these great sea-turtles. How welcome then has been the discovery during the last two years by Mr. Charles Sternberg in the Niobrara Cretaceous of western Kansas, of the nearly complete specimens of *Protostega gigas* which permit the present description of the organization of the limbs, the most important of the parts yet undescribed as well as the very least likely to be recovered in complete form." (Fig. 21.)

This rare fossil was briefly mentioned by Professor Osborn also in *Science* as a " complete skeleton of *Protostega* which lay on its dorsal surface with fore limbs stretched out at right angles to the median line of the carapace, measuring six feet between the ungual phalanges."

A second specimen, which I discovered and sold directly to Dr. W. J. Holland, the director of the Carnegie Museum, is thus described by Dr. Wieland on page 282 of the Memoirs, under the heading "Specimen No. 1421, Carnegie Museum Catalogue of Vertebrates":

"This fine fossil is from the Niobrara Cretaceous of Hackberry Creek." (I should like to correct this mistake. It was found about three miles northwest of Monument Rocks in a ravine that empties into the Smoky, east of where Elkader once stood.) "The *ex situ* portions of the original skeleton, which had weathered out and are secured in more or less complete condition, include the left humerus, radius, ulna, etc. The *in situ* portion consists of the right anterior part of the skeleton, and was secured on a single slab of matrix, in which it still remains intact, as shown in the accompanying drawing by Mr. Prentice, including the lower jaw in oblique inferior view, the skull, the T-shaped nuchal (plate) and two marginals. It will be seen what exceedingly satisfactory information is furnished by the present specimen as compared with all other examples of *Protostega* hitherto found. Specimen 1420 [my first specimen] is more complete than any other at present discovered. As originally embedded in its matrix of chalk, nearly every element was present in an exactly or approximately natural position. Unfortunately, the collector of this surprisingly complete fossil, in an attempt to remove and separate the bones from their original matrix of chalk, mismarked some of them, and also made it impossible to either replace more than a few of the marginal elements, or to determine the outlines of any of the plastral elements. Such work is difficult enough in well-equipped laboratories. However, none of the bones of the limbs are broken, and Mr. Sternberg redeemed himself by discovering and securing in such excellent condition No. 1421, as just related."

I learn from one of the Museum's staff that this specimen is to be mounted this summer of 1908, and placed on exhibition. As long as the Carnegie Museum stands, this splendid example of the great sea-tortoise will be admired by lovers of nature. In shape it is very like the present-day turtle of the Mediterranean. Its huge front paddles, with a span of ten feet, were armed with horrid claws. The hind ones were stretched out parallel with the body and used as sculls by this "boatman of the Cretaceous."

An account of my work in the Kansas Chalk would not be complete without some mention of my discovery, in several small localities, of the crinoid *Uintacrinus socialis* Grinell. According to Mr. Frank Springer, our noted American authority on this subject, only seven localities were known in 1901; he did not know of my discoveries. I can bear witness with him, though, to the rarity of this species.

During the fifteen years in which I have gone over the chalk exposures again and again, I can remember only three localities of these fossils, the Martin locality, another three miles to the east of it, and a third on Butte Creek near Elkader. The first has yielded the finest specimens among those which were described by Mr. Springer in his magnificent treatise on *Uintacrinus*, published by the Museum of Comparative Zoology at Harvard University.

Last year, however, my son George found two splendid specimens about fifty feet apart, further east than they had been discovered before. The locality is south of Quinter, in the southern part of Gove County, thirty-seven miles east of the Martin locality. These two colonies each contained about forty calices. As usual, they are flattened out on the under side of a calcareous slab about a quarter of an inch thick and beveled off as thin as paper at the margins. One slab was sent to the Senckenberg Museum in Germany, while Mr. Springer secured the other.

The calyx, or as we have called it, " the head," has ten long arms, some of them about thirty inches long.*

These beautiful globular animals were stemless, and evidently lived in swarms, as single specimens

* A restoration of the *Uintacrinus* is shown in the same illustration (Fig. 11*a*) in which the *Clidastes* is represented.

are never found. According to Mr. Springer, when death overtook one of these swarms, it fell to the bottom, where the first individuals were buried in the soft mud and preserved, while the others, not being so protected, disintegrated. The limy plates of the calices and those of the arms, which were thus mingled together above the perfect specimens, became compressed into a hard slab, in the bottom of which the perfect specimens are firmly impressed.

Great numbers of these creatures have been discovered in the English chalk, but they consist only of the disintegrated plates.

CHAPTER V

DISCOVERY OF THE LOUP FORK BEDS OF KANSAS AND SUBSEQUENT WORK THERE, 1877 AND 1882-84

BOUT the first of July, 1877, I received orders to go north to the Loup Fork River in Nebraska to search for vertebrate fossils in beds of the Upper Miocene, called by Hayden the Loup Fork Group. I happened to meet, however, an old line hunter, Abernathy by name, who had brought into Buffalo his last load of buffalo hides, and he told me that a little above his cabin, on the middle branch of Sappa Creek in Decatur County, there was the skull of a mastodon, sticking out of the solid rock.

As a visit to his house would not take me far out of my way, I followed his lead; and thanks to the observation of this old hunter, who was scalped in front of his door the next year by a band of hostile Kiowas, I had the privilege of discovering the rich fossil beds of the Loup Fork Group in northwestern Kansas, and found enough to do without crossing into Nebraska.

The whole country north of Buffalo was without human habitation until we reached the old man's cabin. On our way there, as we were driving one sultry day down the long slope to the south branch of the Soloman, we chanced to look behind us, and as high as the eye could reach, the air was as black as midnight with flying dust, dry grass, and buffalo chips. Experience had taught us what all this meant. Will Brouse laid the whip to the ponies, but they did not need it. They, too, had taken fright, and tore down the hill at breakneck speed. On reaching the valley, we came upon a perpendicular bluff, over twenty feet high, impinging on the level flat, and Will swung the horses under its protecting shelter. We sprang out, and while one of us unhitched and tied the horses, the rest caught hold of the wagon and held it down. In an instant all was dark, while the rush of a mighty wind swept over us with a terrible roar and passed on, leaving a calm in its wake. As we followed its trail along the river, we found large trees twisted off at the stump or broken to pieces, their branches scattered like straws.

About sundown one evening, the old man pointed out, in a side draw of the middle fork of the Sappa, his mastodon. I sprang from the wagon, shouting, "It's a monster turtle!" And so it proved to be, a great land turtle, over thirty inches long, twenty-eight inches wide, and fifteen inches high; *Testudo orthopygia* Cope called it. The back of the carapace was sticking out of a ledge of grey sandstone. We applied our picks, and soon had the specimen collected. (Fig. 22.)

Now began an extremely interesting search for this new fauna in Kansas. The rocks in this part of the state usually consist of gray sand cemented together with washed chalk and soluble silica. The foundation on which these beds were deposited is the Niobrara Group of the Cretaceous. The river beds were cut in this soft lime, and later on the wash of the land mingled the whiting with the sand and gravel which the streams brought down from the mountains. The tops of the hills are capped with this conglomerate gray sandstone in ledges many feet in thickness, and as the materials composing it easily disintegrate, great masses of it lie at the bases of the cliffs, resembling old mortar. I called them mortar beds, and the stratigraphers have adopted the name. Indeed, they are mortar beds not only in name, from a fancied resemblance to mortar, but in fact, as all the early settlers can testify. It was no trouble for them to find beds so soft that the material could easily be dug out, and when mixed with water and spread with trowels over the inside walls of a sod house, it made a very comfortable home. When it comes to comfort, the settlers of the short-grass

FIG. 22.—FOSSIL SHELL OF GIANT LAND TURTLE, *Testudo orthopygia.* Discovered by Charles Sternberg in Phillips Co., Kansas.

FIG. 23.—THE SNAKE-NECKED ELASMOSAURUS, *Elasmosaurus platyurus*. Discovered in the Niobrara Group of the Cretaceous. Restoration by Osborn and Knight. (From painting in American Museum of Natural History.)

country have gained nothing by building frame instead of sod houses. The early settler's sod house was cool in summer and warm in winter, and those who live in more modern houses in order to keep up with the times will even now speak with regret of the change.

Not only did I secure a number of specimens of these great turtles, so abundant at this time, but also large quantities of the remains of a rhinoceros. Cope thought it hornless, and named it *Aphelops megalodus,* but since then Hatcher has found that the male bore a loose horn on the end of the nasal bones.

I also got specimens of the great inferior tusked mastodon, *Trilophodon campester* Cope. This remarkably primitive mastodon had a lower jaw that projected beyond the molar teeth for two feet in a straight line, with a socket on either side, containing two powerful tusks that terminated in chisel points. One specimen, which I discovered in 1882 for the Museum of Comparative Zoology in Cambridge, had a jaw four feet long, including the tusks, which extended eighteen inches beyond the end of the jaw.

A set of jaws was brought me by my son last fall. It belongs to a new form of this gigantic pachyderm, which, during the Loup Fork times, inhabited northwestern Kansas and a vast territory west and northwest as far as the John Day basin in eastern Oregon. A remarkable peculiarity of this specimen is that the symphysis is greatly elongated and curves downward thirteen inches below the level of the alveolus, which bears the great molar teeth. This individual was an old animal, as he had shed his first dentition and all the premolars and molars of the second except the very last, those which we call wisdom teeth. Even these are well worn; so the days of the mastodon's life must have been numbered even if he had escaped his enemy, the great saber-toothed tiger, which preyed on him and the other herbivorous animals of the day.

The length of these remarkable jaws is four feet and one inch. The height at the condyle, where they connect with the skull, is thirteen and a half inches; length of molar, nine and a quarter inches; height of crown, two and one-half inches; distance between the two molars, four inches. The sockets for the great inferior tusks are two feet long and six inches in diameter, and the huge recurved tusks themselves must have been over four feet long. Only a sight of these peculiar jaws, with tusks above and below, can give the reader an idea of the formidable appearance of this early mastodon. By the large size and downward curvature of the lower tusks, this mastodon suggests the great *Dinotherium* of the Lower Pliocene of Europe. I regret for America's sake, but I am glad for the sake of the world, that these jaws of the largest mammal ever found in Kansas will find their last resting-place in the great British Museum, where many of my finest discoveries have gone.

Another splendid set of lower jaws I found in 1905 in the Sternberg Quarry, of which I shall speak later, for the Royal Museum of Munich, Bavaria. Part of the symphysis was broken off, as were also the inferior tusks. The length of the jaw as preserved is two feet, six inches and a half, and the height of the condyle, fourteen inches. In the center of the grinding surface, the height is nine and a half inches. The length of the molar is about seven and a half inches, and the width three and a half. This is Professor Cope's *Trilophodon.*

We found near this mastodon many chisel-like tusks that had fallen out of their respective jaws and lay scattered with the other bones. By comparing this specimen with the new species, it will be noticed that there is quite a difference in size, though evidently they were about the same age, as in both cases all the teeth have been discarded except the last molars.

The teeth of these animals were kept sharp by the sand that adhered to the roots on which they lived. Falling into the pits and valleys between the crests of enamel, it scoured away the dentine and cementum, and kept the great grinders ever sharp and ready for use. It is a distinguishing characteristic of these early mastodons that their tusks have a strip of enamel along the inside, while the modern elephants' tusks have only a vestige of enamel at the extreme tip that is quickly worn off.

Another remarkable inhabitant of Kansas during the Loup Fork Period was the three-toed horse, an animal but little larger than the newborn colt of an ordinary farm horse, which evidently lived in herds, judging from the great quantity of loose teeth that we have found. Its toes were spreading, which enabled it to walk over bogs and mossy quagmires on the shores of lakes or rivers, and thus escape the fangs of bloodthirsty tigers by venturing farther out on the soft ground than they dared to follow.

In 1882, while employed by the Agassiz Museum, I found the famous Sternberg Quarry at Long Island on Prairie Dog Creek in Phillips County. I had been exploring for weeks the region at the head of the branches of Deer Creek, which spread out in the divide like a fan; but although once in a while, especially in the neighborhood of Bread Bowl Mound, I had found fragments of the bones of Loup Fork animals in the sod, I had not met with much success, as the rocks here disintegrate so easily and hold moisture so readily that the whole country is covered with grass. There are thirty-three streams in this county as a result of the immense amount of moisture which accumulates in these sandstone beds and is carried to the surface in springs.

One very hot day I started to cross the divide to Prairie Dog Creek. I had the wagon sheet stretched over the bows, the sides lifted to admit the breeze, and sleepy with the heat, I let the horses go on about as they pleased; not noticing, until the level rays of the sun warned me that it was time to camp, that I had gone farther east than I had intended. I had my camp outfit with me, however, and as I saw a bunch of trees in a ravine a mile from the creek I knew that there must be water there. So the three requisites, grass, wood, and water, were at hand.

After pitching the tent, and starting supper, I found to my delight a large exposure of hard siliceous rock, consisting of sand and chalk held firmly together by soluble sand, which proved to be the bottom ledge of a deposit of gray sandstone. I soon found above it a mastodon's bones. My joy knew no bounds, however, when following the narrow draw up to its head, I found that it cut through a quarry of rhinoceros bones, which were sticking out of the sand on either side, while the narrow ditch at the bottom was filled with toe bones, complete or in fragments, and broken skulls and teeth without number. I have collected fossil vertebrates and plants since I was seventeen years old, but this is the greatest deposit of fossils that I have ever discovered.

I shall never forget how, carried away with enthusiasm, I took possession in the name of Science of the largest bone bed in Kansas. I did not stop to ask whether anyone else had any interest in the land, nor did I think it necessary. I had grown so used in my own case to putting aside every other consideration for the sake of the advancement of science that it did not occur to me that anyone else might take a different view. But one day, as I was working in the ravine, an old man, plowing corn, drove up to its eastern edge. When he made the turn, he chanced to look across and saw me, pick in hand, diligently uncovering the skull of a rhinoceros from the sandbank on the other side. He instantly shouted with all the strength of his lungs, "What are you doing?"

"Digging up antediluvian relics," I shouted back. We both shouted as if we were a hundred yards apart.

"Well," he called, "get out of there!"

"All right," I answered in the same loud tones, and kept on working.

The old man, whose name I learned later was Mr. Overton, disappeared, and I heard no more of him until I went into Long Island for food, or grub as they say in the West, and was told that he had come in to a justice of the peace and asked for a warrant to arrest me for collecting these old bones. He never again came directly to me, either that year or the following, but people told me that he went around to all the justices in that part of the country, trying to get his warrant. Finally, however, they managed to convince him that I was not harming him, and was benefiting science.

Two years later, in 1884, I was employed by the late Professor Marsh to explore this same fossil bed. The bones which I was after now were covered by fourteen feet of moulding sand and a four-foot ledge of hard rock, the heavier bones lying on the sandstone, the lighter ones mingled with the sand above. This sand and rock had to be removed by pick and scraper, which meant that there was a large amount of heavy labor before us. Therefore, having more means at my command than I had had before, I drove up to Mr. Overton's door and offered him forty dollars a month to work for us with his team during the whole summer, with the understanding that I was to have all the fossils found. This offer he gladly accepted, and I found him a very careful worker. Not only did he do the rough work well, but when we got a floor laid bare above the bones, he proved to be a most careful collector. My other assistant on this expedition was a Mr. Will Russ, who afterwards became a skilful dentist.

Our method of work was first to cut down and remove the sand and rock for a space twenty feet wide and perhaps a hundred long, using a plow and scraper. Then we cleaned up our floor and uncovered the bones with oyster knives and other tools which we had made to suit our purpose. One, I remember, was a hoe straightened out at the shank and cut off at the corners to make a diamond-shaped tool. With this we could work under the high bank, and take out specimens which we could not reach otherwise. Trowels and diggers of various patterns were used also.

The bones which we were collecting lay scattered along both sides of the ravine for a quarter of a mile, often in pockets or pot-holes in the gray sandstone. Of this there are two layers, about fourteen feet apart, the interspace being filled with beds of fine moulding sand, with some whiting from the underlying chalk, which constituted the land surface when these fresh-water beds were deposited. There are also beds of sand that have been washed clean by the currents of the flood-plain of some ancient river, for the exposed section shows all the different deposits of an overflowed valley. Above the washed sand is a stratum of sand and clay, indicating that here was a quiet place where the muddy backwater deposited its load. This layer, upon exposure, cracks in all directions, like the mud at the bottom of a puddle after the water has evaporated.

It has always been a problem to account for the number of the animals represented here, and for the fact that the bones are so scattered. All parts of the skeletons are mingled in the greatest confusion, with no two bones in a natural position. One is, of course, forced, after an observation of this country, to agree with Drs. Matthew and Hatcher that these bones were deposited in the flood-plain of a running stream and not in great lakes, as was believed by older geologists. But the only supposition upon which I can account for the intermingling of all the bones of the skeletons on the bottom sandstone layer is that the fine sand through which the bones were distributed, becoming saturated with water, was converted into a quicksand, in which the bones sank until they reached the impenetrable layer below; the heavier bones of course being at the bottom.

What caused the death of the countless individuals in the Sternberg Quarry, is a question not easily answered. The authorities quoted above believe that during the Upper Miocene Period, there were many water-courses separated, by slightly elevated divides and broad flood-plains, with possibly here and there small lakes, where the dense vegetation had clogged some sluggish stream. But during a rainy season of unusual duration, the whole region for many miles must have been converted into a series of lakes; and all the animals in the vicinity, after having gathered at the highest points they could find to escape death, must have been finally overwhelmed by some great flood that covered every inch of ground. Then after maceration took place, the bones might have been scattered by other floods.

A theory of my own, equally plausible, is that the animals were buried beneath a sandstorm, which tore loose the fine sand of the flood-plain, and scattered it in suffocating volumes over the frightened multitudes which had herded together in search of safety or courage.

This land, now three thousand feet above sea level, was only a few feet above when these rhinoceroses moved over it in countless herds. Everywhere were swamps filled with sponge moss, and tropical streams, whose wealth of vegetation formed thick jungles along their banks. On firmer ground, great areas were covered with a dense growth of rushes, through which the paths of these animals were the only trails; while higher up still, the soft damp soil gave a foothold to forests, through which the great mastodons sounded their trumpet calls, as they roamed about, tearing up trees with their powerful trunks and feasting upon the rich, juicy roots.

That year, 1884, in which I explored the quarry at Long Island, was a memorable one, not only be-

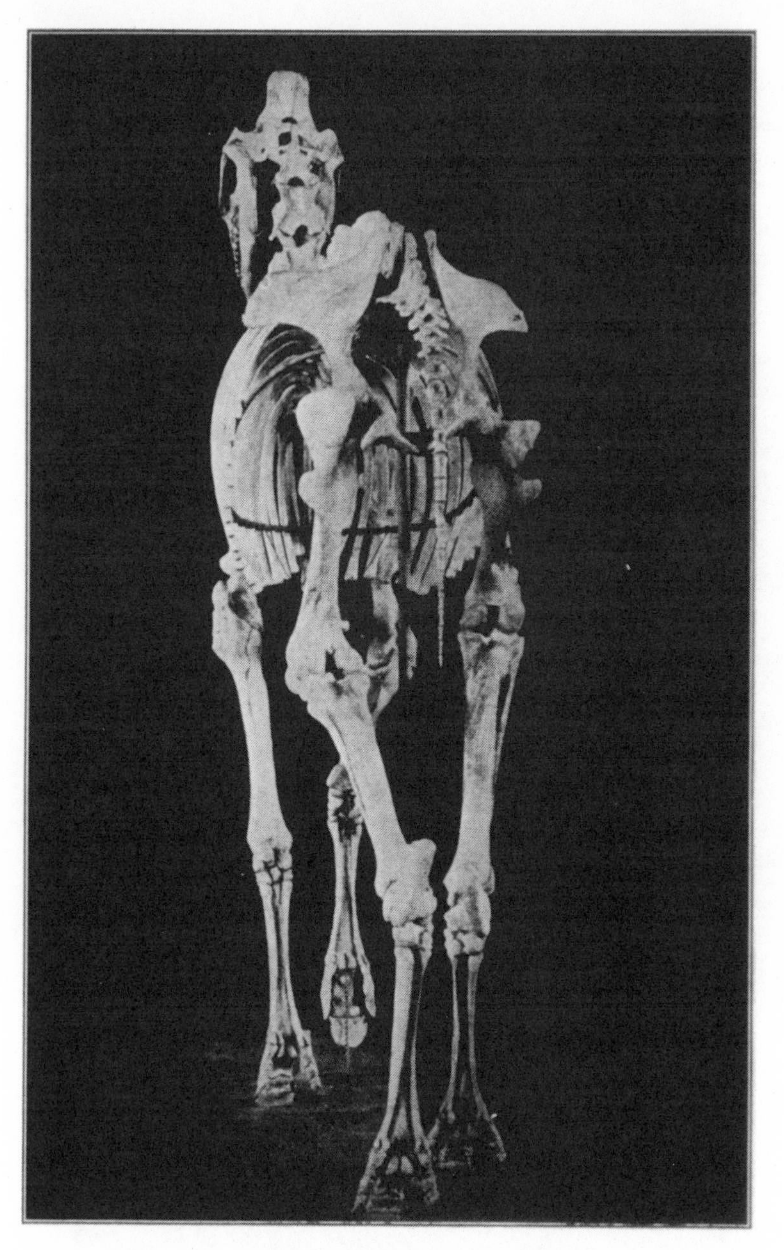

FIG. 24.—THREE-TOED HORSE, *Hypohippus.*
From the Middle Eocene of Colorado. (After Gidley.) In American Museum of Natural History.

FIG. 25.—FOSSIL RHINOCEROS, *Teleoceras fossiger*.
From Sternberg's quarry at Long Island, Phillips Co., Kansas. Collected by Wortman; mounted in the American Museum of Natural History. (After Osborn.)

cause we secured a large carload of rhinoceros bones, but also because we had with us Mr. J. B. Hatcher, who afterwards helped to build up three great museums of vertebrate paleontology,—the museums of Yale and Princeton and the Carnegie Museum. With the last he was connected at the time of his death in 1904, just twenty years after he made his first collection of vertebrate fossils with me. A bright, earnest student, he gave promise of a future even then by his perfect understanding of the work in hand and the thoughtful care which he devoted to it. I have always been glad that I had the honor of being his first teacher in the practical work of collecting, although he soon graduated from my department, and requested me to let him take one side of the ravine while I worked the other. He employed Mr. Overton's son with a plow and scraper, and got out a magnificent collection with no further instructions from me.

That same year Professor Marsh came to my quarry and leased it from the owner, and I never saw it again until 1905, when I came into my own once more, and in addition to the splendid mastodon, mentioned earlier in this chapter, found the material for two perfect mounts of the rhinoceros. One is to be mounted at Munich, the other at Bonn.

With Professor Osborn's consent, I give a photograph of the fine specimen (Fig. 25) which Dr.

Wortman secured in 1894 from this quarry for the American Museum. A vast collection from the same spot is stored in the National Museum in its original packages, with which I filled a car in 1884. I saw there a whole case filled with the skulls of the rhinoceros *Teleoceras fossiger,* which I secured in great numbers at Long Island.

It is strange to think that the foundation on which these beds of fresh-water deposits lie unconformably is the great Cretaceous sea bottom, whose tilted and uplifted strata tower two thousand feet above the carboniferous rocks in eastern Kansas. The Republican, Smoky Hill, and Kansas rivers have carved their way through all these strata, so that by following down these streams, one can get cross sections of the country.

I have often asked men who were sure that there must be coal beneath the surface, why, instead of hiring a man to dig a hole for them, they did not hitch up their buggies and follow the valley of the Smoky Hill, beginning at the Colorado line. The first stratum exposed is of course the recent, with its sandy loam; in it, here and there, a crumbling buffalo skull or an eroded implement. Then comes the Pleistocene deposit, consisting of clay, sand, and fragments of rock mingled together. From this formation I secured over two hundred teeth of the great Columbian Mammoth. Next come beds of black shale with giant septaria, the Fort Pierre Group of the Cretaceous, whose upper beds we explored in Montana in 1876 for dinosaurs. In this formation, in Kansas, I found a new species of *Clidastes*. The specimens are now in the Kansas University collection, and the species has been named by Dr. Williston *Clidastes westi*, in honor of the Kansas University collector, the late Judge E. P. West.

We have not gone far down the river below the forks, before this formation, which at McAllister topped the hills, passes under the river. Then reddish and blue chalks occupy the country for some miles, and in turn disappear to give place to yellowish and blue chalks, which finally make way for the blue and almost white chalks that run under the river near the mouth of Hackberry Creek in eastern Gove County.

At White Rock in Trego County the hard white limestone, in fortification blocks, is piled ninety feet high. Further down appears the post limestone of the Fort Benton Group, with its characteristic *Inoceramus* shells; while in central Kansas, brown and white sandstone and brilliantly colored clays occupy the whole region for sixty miles, giving place at last to the hard limestones and the friable shales and sandstones of the Upper Carboniferous. No coal, except very shallow veins in the Upper Carboniferous and the Dakota Group of the Cretaceous, has even been found in this big ditch, which, less than a quarter of a mile wide at the head of the Smoky Hill branch at Wallace, broadens out to a width of several miles at the mouth of the Kansas River.

It is impossible to compute the vast amount of mineral matter which has been cut out from these Kansas plains and carried by the river into the Mississippi and on to the Gulf. Since the first narrow trench cut its way through the hardened ooze of the Cretaceous ocean bed, all the flood-plains of the Missouri and the Mississippi below Kansas City have been enriched by the material that once covered these valleys of Kansas, and the delta below New Orleans has been partly built up by it.

It may interest my readers and give them a glimpse into the daily routine of a fossil hunter's life, if I quote one or two notes from a diary which I kept during my work in these Loup Fork beds.

"Friday, July 11.—This is to record the most successful day since we have been in the field. We have collected three sets of under-jaws, three skulls. It has been extremely hot. We have put in eight hours of hard work."

"Saturday, July 12.—To-day I got out and packed our three skulls and three lower jaws. They were within the space of a square yard. We got some very fine bones, and best of all, a perfect front foot in position, a perfect humerus, a perfect femur, except proximal articulation, the premaxilla of a cat with a huge canine (saber-toothed tiger). We got great quantities of the bones of the feet, an axis, and one other vertebra in good state of preservation, a fine scapula, etc. This afternoon has been the hottest day of the season, but this evening the wind changed to the north, and it is quite cool. I got in addition to the specimens mentioned a maxilla of a saber-toothed tiger. The enormous young canine was two inches long and three-quarters of an inch wide."

I might go on and quote indefinitely, but the story would be about the same. I recall, however, one or two incidents connected with my work in this field, which may be amusing or interesting to my readers.

Once in 1882, while collecting for the Museum of Comparative Zoology of Harvard University, I met an old gentleman and his dear old wife, the hair of both showing upon it the snows of many winters, sitting on a board laid across a dry-goods box to which two wagon wheels had been attached. A team of ponies harnessed with rope instead of leather, with lines of the same material, completed the outfit. The old man and his wife sat up very straight and dignified and demanded of me what I was doing in that part of the country.

"Oh," I answered, "I'm looking for rhinoceros bones in the loose sand of the hills here."

"Well," the old man said, "I am interested in these old bones myself. I don't claim to be a scholar; in fact, I am quite illiterate, but I think when this earth was in a molten state, these old hippopotamuses wallowed around in the mud and got congealed in the rocks."

The following incident I did not find quite so amusing. One day I discovered turtle shells sticking out on either side of a narrow gulch which cut through a large deposit of sand. In digging out those already in sight, I found many more; collecting in all some twenty fine specimens, but all quite small. Following down the gorge, I discovered that it opened out, on Beaver Creek in Rawlins County, into a great amphitheater several acres in extent and almost denuded of vegetation; an ideal place for fossil hunting, as the elements had been digging out and removing the sand for ages. And sure enough, I soon stumbled upon the complete shell and skeleton, four feet in diameter, of a specimen of Cope's *Testudo orthopygia;* but it nearly broke my heart to find that while the specimen had weathered out in a perfect condition, some vandal—for I shall ever maintain that the wanton destruction of life that now is or of the remains of life that once was, is wicked,—some man had chopped it all to pieces with a mattock.

Passing on in a not very pleasant frame of mind, I came upon another individual of huge proportions, which had suffered the same fate, and then upon another; all that this rich-looking ground afforded had been utterly ruined.

Angry at the thought that any man should commit such sacrilege,—for to me these footsteps of the Creator in the sands of time are sacred,—and bitterly disappointed, since I knew that I should very likely never again come upon such huge specimens of the reptilian life of that age, I walked into camp blinded by hot tears, and failed to notice a stranger who was sitting there on a box.

"Some infernal vandal has been up this ravine," I shouted to Will, "and dug up with a mattock three of the finest turtles I ever saw."

As if he had been shot, the man jumped from the box and exclaimed in accents of heartfelt contrition, "It was me. I was out here digging roots to build a fire with, and ran across them. I didn't know they had any value, and I wanted to see what was inside of them and dug into them."

His surprise and dismay were so comical that the murder vanished from my heart, and overwrought as I was, I broke out into a fit of uncontrollable laughter which used me up for the rest of the day.

Another time I had a rather unusual experience. My assistant, a Mr. Wright, and I were digging out rhinoceros bones on Sappa Creek. We had noticed a house on the other side of the creek, although dense timber cut off most of its surroundings, and happening to look toward it once, we saw a girl of about sixteen years rush out from the timber and begin to climb the steep hill toward us. I never saw anyone run so fast up so steep a hill. Her strength failed her, however, when she got to us, and it was some time before she could tell her story. It seems that her mother had gone out to milk, and as the ground was slippery from a rain of the night before, she had fallen and dislocated one of the bones in the palm of her hand.

All the men were away and had taken all the horses, and it was seventeen miles to the nearest doctor. The girl, knowing that we were digging up bones, had concluded that we could set them, and had come to us for help. Although I had never attempted anything of the kind before, I could not resist the poor child's appeal and went to the house. The mother lay moaning on her bed, and would answer nothing when I asked whether I should try to set her hand. But as the girl was very desirous that I should make the attempt, I decided to do so.

So while Mr. Wright held the arm, I put splints and a roller bandage under the hand, which was laid on a table, and then forcibly pushed the bone back into its natural position. After which I bandaged the hand tightly. I left directions with the girl to hang a can of water with a small hole in it over the hand, so that the water might drip on it and by evaporation cool it and prevent inflammation. My instructions were carried out by the brave girl, and her mother's hand was soon as well as ever.

In these last chapters I have often wandered far afield, for it would have taken too long to relate all the events of my various expeditions in consecutive order. Hoping that my readers will pardon the digressions, I return to the expedition of 1877.

Russell Hill proved a most efficient assistant, and it has always grieved me that he should in later years have given up work in the fossil fields for the practice of medicine. Will Brouse, too, was an enthusiastic worker; he was not satisfied to be relegated to the pots and kettles and horses, and not only did his duty as our teamster and cook, but soon accomplished almost, if not quite as much in the field as any one of us. I never had a more congenial party in all the years of my field work.

But one day in August I received a bulky letter from Professor Cope. "Turn over all the outfit to Mr. Hill," he wrote, "and go at once to a new field discovered in the desert of eastern Oregon. Go to Fort Klamath, Oregon, and from there to Silver Lake, to a man by the name of Duncan, the postmaster. He will guide you to the fossil bed in the heart of the sage-brush desert. You will likely find human implements mingled with extinct animals. You are to go secretly; tell no one where you are going. Have your mail sent by a circuitous route, so you cannot be traced."

I received the Professor's order with excitement and great joy; but in spite of his injunction to start at once and without communicating my intention to anyone, I could not bring myself to leave for the Pacific Coast, to be gone for an indefinite time, without bidding good-by to my father and mother, and I concluded that even if someone should find out where I was going and try to follow me I could easily give him the slip and get to the field first.

Buffalo, the nearest railway station, was seventy-five miles away, a two days' journey, with our big load of fossils. So I mounted my riding pony and made the long trip the next day, reaching the station at sunset, tired and sore. My pony, however, endowed with the enduring power characteristic of a good Indian pony, was still fresh enough to shy at a rattlesnake in the road, and as I happened to be sitting sideways in the saddle, throw me to the ground within a few feet of the snake.

That night I went to my home in Ellsworth County, bade my dear ones good-by for an indefinite length of time, and was back at Buffalo again at midnight of the following day. My boys met me at the station with my roll of blankets, tools, and baggage, and away I went to " fresh fields and pastures new."

CHAPTER VI

EXPEDITION TO THE OREGON DESERT IN 1877

T Monument Station, I was surprised to see Mr. S. W. Williston get aboard with all his outfit. Williston did not know at first that I was on the train, and when he entered my car, he was greatly astonished, thinking that I was on his trail. He tried to find out my destination, but failed. We slept together at Denver. Then he took a train south, while I went north toward Cheyenne and the West.

Onward our train sped toward the land of the setting sun, through the grand and impressive scenery of the Rockies and Sierra Nevadas. At Sacramento I took the railroad for Redding, where, with seven other passengers, I entered a Concord coach drawn by a team of eight horses, and continued my journey by stage.

It was a lovely August evening. The moon was at its full, and the night was almost as bright as day. No sound broke the deep silence, except now and then the whoo of an owl as it called to its mate far away in the depth of the forest, or the plash of running water falling in cascades over the shelving rocks and dashing against the boulders.

Higher and higher we climbed, through primeval forests of spruce and fir, whose branches clove the sky a hundred feet above our heads. The rarefied air filled our lungs with its life-giving tonic, exhilarating us like wine. We knew that far above us rose Mount Shasta, the giant of the range, but for a time the heavy timber shut out the view, and we could see only the road ahead, winding up and up through the forests. Then suddenly, without warning, we moved above the timber-line, and Mount Shasta stood revealed in all its beauty, a perfect cone, towering four thousand feet into the air, its robes of everlasting snow glistening in the moonlight. Above, in the clear blue of the sky, the stars sparkled like jewels in an immortal canopy.

It was the first time that any of us had looked upon that majestic scene, and whatever may have been the differences of temperament among us, we were one in the feeling of awe which the glorious picture inspired. It laid a spell upon us; we were dumb before the invisible presence of the Power that had reared this stupendous pinnacle, and involuntarily our thoughts turned to that "city that hath foundations, whose builder and maker is God."

Then to break the awful silence, and give some

vent to our emotions, we broke out into the old song, " 'Way down upon the Suwanee River"; and so we journeyed on for many hours, never out of sight of that majestic form.

At Ashland I was obliged to wait for a driver with a buckboard and a team of ponies to take me to Fort Klamath, Oregon. I was at that time a great lover of the gentle art of fishing, and early in the morning, before it was fully light, I was astir among the great live-oaks that grace the town. Walking through the sleeping village, I ran across the footprints of a large grizzly bear in the dust of the road, and followed them through the vacant streets. Wherever a gate had been left open, the bear had entered the yard, walked around the house, and come out at the gate again. I hoped to get a glimpse of him, but was disappointed, as the tracks led into the gloom of the forest. So I went fishing, and caught some speckled beauties for breakfast.

That evening I was driven over to Fort Klamath, where I was kindly invited to take possession of the commanding officer's quarters and make myself at home; an invitation which I proceeded to accept at once.

Learning that a sheep-owner a few miles away had killed a grizzly, I went out to his camp to see it. Sure enough, there lay the mighty carcass, encircled with four inches of grease, enough for the polls of all the boys in Oregon. It seemed that as the time for his winter nap was approaching, Mr. Bruin had been laying in a supply of fuel by devouring the fat wethers of our friend's flock. The latter had built a heavy brush fence around the sheep, and with the help of a large number of hounds, had kept his range free from coyotes, but he had been helpless before the attacks of this big bear. When he watched on top of the brush fence, he was not molested, but no sooner did he seek the comfortable cot in his tent, than his slumbers were broken by the piteous bleat of some sheep, as it was carried off to the woods by the bear.

About ten days before I reached Klamath, he had been awakened in the middle of the night by a commotion in the flock, and rushing out in his shirt into the cool night air, had seen the bear only ten feet away, across a deep and narrow stream. Without thinking of the consequences to himself if he only wounded the creature, he opened fire with his Winchester, and the first shot broke the bear's neck.

When I arrived, the skin had been removed, but the huge carcass, which must have weighed at least a ton, had been lying in the hot August sun ever since. The sheep-owner (I am sorry that I have forgotten his name, as I was under heavy obligations to him) promised me that after breakfast he would help me in the not very enviable task of removing the decaying flesh from the bones. But after one whiff from the windward side, he asked a pertinent question, was I fond of trout, and upon my answering yes, remarked that he knew of a creek where he could get some beauties, and immediately disappeared. I saw him no more that morning.

At the first thrust of my knife into the bear, the stench was so horrible that I grew deathly sick. I filled my pipe and tried to find relief in smoking, but even then the odor was overpowering, and I smoked and sickened through the livelong day, until I had cleaned the filthy flesh from the bones, and they had been tied up in gunny-sacks and hung in a tree to dry. Then into the creek I went and with soap and sand scrubbed and scoured my body; but the horrid smell still hung about me, and I could eat neither supper nor breakfast the next morning, although at dinner I managed to stow away a good square meal. But even now, after thirty years, if you say "bear" to me, I can smell that bear.

At Klamath I hired for my assistant a man named George Loosely. I also bought two saddle ponies and one to carry the pack; and with a government tent and other outfit and rations purchased at the commissary,—we had our flour baked into bread by the post baker,—we started for Silver Lake, although no one at the post could give us any directions. I had a department map, sent to me by Professor Cope, which recorded, mistakenly as we found later, that Sprague River rose in Silver Lake. The government road to the east crossed the Williamson River on a government bridge, and came to an abrupt end in an Indian village on the western bank of Sprague River. So we decided to take the road as far as we could and then follow up the river to its source in the lake.

When we reached the Williamson River, we found there the lodge of a Snake Indian, who appeared dressed in red paint and a breech-cloth, and demanded toll. But as American citizens we had paid taxes to help pay for that bridge; so we refused to pay toll for the use of our own property, and rode across in spite of the threats hurled at us.

We reached Sprague River that same evening, and went into camp a short distance from a large Indian town. The houses, built by government contractors of rough logs, consisted of a single room with a shake roof. The Indians had torn out the board floors, and instead of using the fireplaces and chimneys which the builders had erected for their convenience, they had cut holes in the roofs, and built their fires in the middle of the floor, sleeping around them at night as their fathers used to do in their lodges or Sibley tents.

George, who was more familiar with them than I

was, learned that a chief lay dying in one of the houses, and after supper he left me and went to witness the death ceremonies. After stowing away the bread and coffee between our mattresses and covering them with blankets, and hiding the bacon at the bottom of the mess box with tin dishes piled on top of it so that I should hear the rattle if a thieving Indian attempted to get at it, I, being tired, dropped off to sleep.

About three o'clock in the morning, George appeared, having been shut up in the house with the dying chief all night. When the medicine man began his incantations, the doors and windows were closed, while the steaming Indians danced in a circle around the dying chief, forcing the unwilling George to take part in the ceremonies. All night long they moved around in their death dance to the music of their drums and the wild gesticulations of the medicine man, and when George finally got away, he was about exhausted. He was soon lost in sleep, and as I habitually lie on my sound ear, neither of us heard anything through the night. But the next morning, when George had put on the coffee to boil and went into the mess box for the bacon, it had disappeared. The dishes had been carefully replaced.

After a breakfast of bread and coffee, we were early in the saddle, taking a heavy trail that led north and skirted Sprague River. By the merest chance, we met a white man, the first we had seen since leaving the post, and we stopped to ask the way to Silver Lake. A number of Snake Indians were standing around at the time. The man told us to go north on the trail to a sheep camp in Sican Valley, where we would receive further directions, and thanking him, we rode confidently forward.

Just as the sun was sinking, we entered a splendid forest of fir and spruce, and soon found that our trail forked. The heavy, well-traveled branch turned a little west of north; the other, leading due north, had apparently not been used since last year, as it was covered with old leaves. We did not know what to do, as the man whom we had met in the morning had not mentioned this fork. While we were talking about it, we heard the jingling bells of a pack horse or Indian cayuse, and soon a boy hove in sight, driving a couple of pack ponies. Moving to one side to let him pass, I asked him where he was going.

"To Sican Valley, to a sheep ranch," he answered, and immediately was lost to sight among the giant trees. We meekly fell in behind and hurried after him.

Suddenly we came out into a natural park, the end of our trail. Five Indian lodges stood about in the open space, and five valiant braves, in their usual attire of paint and breech-cloths, with the inevitable Winchester, stepped forward to inform us that "white man was lost in the woods," and that they would show him the trail for two dollars.

"Where is that miserable papoose?" I demanded, but they only grinned and repeated, "We will show you the road for two dollars."

It was my habit, in a crisis of this kind, to smoke, for I regret to say that I was for many years a lover of the soothing weed; so, drawing out of my saddle-bag a pound of fragrant "Lone Jack," I proceeded to fill my pipe and decide upon my further course. Instantly the Indians crowded around me, and dropping the butts of their guns to the ground, pulled out their tobacco pouches, and opening them wide, held them up to be filled, crying in chorus, "Me tobac! Me tobac!"

But the memory of the deceitful boy was still rankling in my mind. I told George to follow me with the pack horse, and deliberately lighting my pipe and filling my lungs with smoke to their utmost capacity, I blew a cloud of it into the faces of the expectant beggars. Then I drove my spurs into my pony's flanks and started off in a mad race against time, as the long shadows warned me only too plainly that the daylight, our only guide now, would soon leave us. I did not look back, but George, who did, saw the Indians, in anger, level their rifles as they shouted to us to stop.

That race with darkness was an exciting one, but just before night overtook us, we reached the trail which we had left to follow the lying Indian boy. In our haste, our bread had been torn from its sack by the outstretched limb of a tree, and was lost. However, we were so thankful to have escaped paying toll to those filthy Snakes, that we cheerfully made our supper of coffee, and sought our blankets.

At the first streak of daylight, after another meal of coffee, we were in our saddles; and we traveled all day, until, just as the sun was setting, we heard the welcome bleat of sheep and saw the herders driving their flocks down the slopes of the neighboring hills to their corrals in Sican Valley. Following them, we soon spied the camp in the heavy timber and smelled the delicious savor of a pot of mutton that was boiling over the fire. And before long, seated at the rude table, we were enjoying to the uttermost the hospitality of the camp.

We had learned on the journey that Sprague River rises in the heart of the mountains, instead of in Silver Lake, and we had crossed the divide between it and the lake before reaching Sican Valley. The next morning our sheepmen directed us on our way; and that same evening we were skirting the lake's lovely shores. Its wide expanse of water put me in mind of my boyhood days on Otsego Lake or the Glimmer-glass.

We soon reached the hospitable home of Mr. Duncan, the postmaster of Silver Lake. He had built a comfortable house of logs, with a large chimney at one end and an old-fashioned fireplace, around which, as the nights were cold, we gathered and talked until far into the night.

Mr. Duncan's family consisted of his wife and daughter, a dear, good girl, who will forgive me, I am sure, if I tell a story at her expense. George and I were sent to bed in a lean-to, and as our bed-room was next to that of the Duncans and the stoppings had fallen out of some of the chinks in the wall between, we could hear everything that was said in their room. In the middle of the night I woke up and heard the old gentleman talking to his wife about their daughter.

"Mother," he said, "I think John will be a good husband for Mary, don't you?"

Before she could answer, Mary, who had a bed at the other end of the parents' room, called out with great energy, "I think so too, father!"

In an instant all was still, while George and I, in our efforts to keep quiet, stuffed the bedclothes into our mouths until we were almost suffocated.

We unloaded our weary pack horse, and the next

day brought our supplies, and loaded them into Mr. Duncan's wagon. Then taking him with us for guide, we started on our long drive to the boneyard, fifty-six miles through the great sage-brush desert of eastern Oregon.

On we journeyed, through what seemed an interminable expanse of sage-brush, greasewood, and sand. The bunches of sage-brush topped conical mounds of sand, whose sides were scoured and polished by the winds that howled in and out through the labyrinth of hills, laden with drifting sand. If one could have gained an elevation above the level of these sandhills, and looked out over the landscape, one would have gazed upon a scene of even greater desolation than that afforded by the parched short-grass plains of western Kansas,—a dreary, monotonous waste of olive green, stretching away north, east, and south, as far as the eye could reach, and shut in on the west by the great ranges of the Sierras, whose flanks, dark below the timber line with heavy forests, were deeply scarred above with glistening white glaciers.

We followed the California road to Oregon, for in those days Oregon was practically an unknown territory, with the exception of the Willamette Valley. And I suppose that it is still so, for that moist, fertile valley differs as widely from the vast semi-desert east of the Cascade Range as the Santa Clara

Valley from the cactus-covered sandhills of southern California.

At night, after a day's journey through sand and sage-brush, we came to a ranch beside an alkaline lake in the very heart of the desert. Here, in a cabin built of logs from the neighboring mountains, lived the hermit of this region, a man named Lee Button. Had it not been that the road passed his door, he would have seen only a hunter now and then, out after the deer which abounded in the desert, or perhaps the cattlemen when in winter they turned their cattle loose in the desert to look out for themselves. On all the neighboring ranches, the cattle were turned into the desert for food and shelter in winter. Here, protected from storms, they fed upon the alkaline grass and sweet sage and upon the thick leaves which fell in handfuls from the greasewood bushes. These cattle had cut innumerable paths at every conceivable angle, and one unaccustomed to the country might easily become confused and lose himself in the labyrinth of trails. There was horror in the thought of being lost in that solitude.

Mr. Duncan put up his horses in the barn of the ranch, which was well stocked with hay and oats, and we picketed our ponies on a flat covered with alkaline grass on the borders of the lake. Then from under a certain post which he knew of, Mr,

Duncan dug up a tin can containing the key of the cabin. Past experience had taught Mr. Button caution. He had gone to California once, after a herd of horses, leaving his door unlocked, and some prowling immigrant had abused his hospitality and robbed his cabin of its store of food and blankets. So now, when he left home, he locked the door and hid the key, giving, however, the secret of its hiding-place to his neighbor, Mr. Duncan.

His cooking utensils, consisting of a camp kettle, a frying pan, a Dutch oven, and a coffee pot, were brought out and cleaned, and the larder searched for food. It was the custom of the country at that day to consider food and shelter free to all. I was offered the next year a house, blankets, flour, and bacon, as much as I could use for nothing, if I wanted to spend the winter on a ranch in eastern Oregon. I was only expected to cut my own wood and cook my own food.

Soon a cheerful fire was blazing on the hearth, and the burning sage-brush was filling the air with that indescribable odor from which one is never free while in the desert. We had traveled through great droves of wild geese along the lake, and as they were so tame that they simply stepped out of our way like barnyard geese, we did not think it worth while to waste ammunition on them. So I set three traps, common steel traps such as are used for catching coons, and strewed oats around them. The next morning I found a brant in one, a magpie in another, and the house cat in the third. We let the cat and the magpie go, and breakfasted on the brant. Our usual fare was bacon, bread, and coffee, and sometimes dried apples. I worked for years in Oregon with no other food, except an occasional deer or mountain sheep.

The next day, trusting entirely to Mr. Duncan's guidance, we pushed on without a trail, winding in and out among the hillocks with no landmarks but the mountains in the west. At sunset, we came out into the open on the shore of a small alkaline lake. "Fossil Lake," I named it at once, and it goes by that name to this day. This pond, as we should call it in old New York, covered only a few acres then, and is now entirely dried up.

"There," shouted Mr. Duncan, as he pointed with his whip to the lake shore, "there is the bone-yard."

I instantly requested him to help George get supper and pitch the tent, and seizing my collecting bag, rushed down to the shore. The clay bottom of the ancient lake had been dried out, and now formed the shore of the remaining water. This old lake bed had once extended over a much larger area, but it had been partially buried beneath large piles of drifting sand. Scattered through the loose sand

and on the clay bed were great numbers of the bones and teeth of reptiles, birds, and mammals, indiscriminately mingled. I had come upon a bone-yard indeed.

I was down on the sand at once, picking up bones and teeth and putting them in piles. No two bones seemed to belong together, and the skulls and arches had been crushed beneath the feet of animals, probably cattle and deer, which had come down to drink at the lake. What pleased me, however, was the fact that scattered among these remains of an earlier day, were arrow-heads and spear-points of polished obsidian, or volcanic glass. I was too much excited then to notice that I did not find a single bone or tooth in its original position in the clay matrix, but that all were loose, detached, and scattered, and that the implements were lying about in the same way.

As Mr. Duncan was to return to the post-office at Silver Lake the next morning, I gathered a cigar-boxful of loose teeth, arrowheads, and spear-points, and packed them to send off to Professor Cope. And that night, by a sage-brush fire, I wrote the letter which he saw fit to publish in the *American Naturalist*, a magazine of which he was the editor, under a title of his own, "Pliocene Man," and signed "E. D. Cope."

For weeks I sifted through my fingers the fine sand of that lake shore, picking out bone after bone.

The only specimen which I found undisturbed in the clay matrix was part of the skull of a hairy mammoth, or *Elephas primigenius.*

Dr. Shufeldt is the author of a valuable memoir on the fossil birds of this region,—" The Fossil Avi-Fauna of the Equus Beds of the Oregon Desert," published by the Philadelphia Academy of Sciences. He worked over the collection made by the late Professor Condon of the Oregon State University, the collection which Professor Cope made a few years after mine, and mine.

In these three collections, he finds five species of grebes, and nine of gulls, of which two species are new to science, Professor Condon being the discoverer of one, while I found the other. Of cormorants, there are two species, one discovered by Cope. One species, quite common among the fossil remains, is now extinct. There is a new swan also, described by Professor Cope, who writes of it: " This swan was discovered by ex-Governor Whitaker of Oregon [who discovered the Fossil Lake locality] in the Pliocene formations of the state. The same bird was afterwards procured by my assistant, Charles H. Sternberg." Altogether there are nineteen species of Anseres, i. e., geese, ducks, swans, etc., of which two are new.

One of my discoveries was a flamingo, which was dedicated to Professor Cope under the title *Phœni-*

copterus copei. Dr. Shufeldt says: "It is a fact of no little interest that a flamingo inhabited the lakes of the Silver Lake region of Oregon during the Pliocene Epoch." The collections include a heron and a couple of coots also. Among the fowl are four grouse, discovered by Cope, and an entirely new genus and species which I had the honor of finding. Of eagles, there are two species. There are also a great horned owl, a blackbird, and a raven.

Among the other fossil remains taken from this region are six genera of fish, a majority of them new, and fifteen species of fossil mammalia, including two llamas, three horses, an elephant, a dog, an otter, a beaver, a mouse, a great sloth, *Mylodon*, as large as a grizzly bear, and other forms.

"Thomas Condon," writes Dr. Shufeldt in his memoir, "was the first scientific man to visit the Fossil Lake region, with the results already stated. Cope and his assistant Charles Sternberg came later, and gathered many hundred bones and bone fragments." And in the preface to his "Tertiary Vertebrata," Vol. III, page xxvii, Professor Cope writes: "The Tertiary formations explored in 1878 were the John Day, Loup Fork, and Equus beds. These were examined by Charles H. Sternberg both in Washington and Oregon; in the former near to Fort Walla Walla, and in the latter, in the desert east of the Sierra Nevada. The basin of an ancient lake, originally discovered by Governor Whitaker of Oregon, was found strewn with the bones of llamas, elephants, horses, sloths, and smaller animals, with birds, and all were collected by Mr. Sternberg and safely forwarded to Philadelphia. I examined this locality myself in 1879 and obtained further remains of extinct and recent species of mammalia found mingled with numerous worked flints."

The reader will notice that Cope puts my expedition in '78 instead of '77 and that Dr. Shufeldt gives Cope's visit to Fossil Lake as before mine, when, in reality, it was two years later.

On p. 420 of his memoir, Dr. Shufeldt writes: "We must believe that it still remains problematical whether man was there, and further comparative search is demanded to decide whence came, and at what time, those stone implements of human manufacture, commingled as they are with the bones of the animals, many of which are long since extinct." And Professor Cope says on the same subject: "Scattered everywhere in the deposit were obsidian implements of human manufacture. Some of these were of inferior workmanship, and many of them covered with a patin of no great thickness, which completely replaced the luster of the surface. Other specimens were bright as when first made. The abundance of these flints was remarkable, and suggested that they may have been shot at the game, both winged and otherwise, that in former times frequented the lake."

After I had written the letter already mentioned, having carefully gone over all the ground in the vicinity of Fossil Lake, and longing for new worlds to conquer, I started out one day on my pony through the desert, hoping to find another locality in which the wind had uncovered a fossil bed. I spent the greater part of the day in fruitless search, and was about to return home when I was attracted by the top of a dead spruce tree sticking out of a sand-hill. The rest of the tree had been completely buried by the sand.

My curiosity was aroused, and I climbed to the top of the hill to examine the spruce. When I reached the top, however, I found myself looking down into a pleasant little valley, which had been scooped out by the wind, and, descending, I discovered that I had stumbled upon the former site of an Indian village. Places near where the lodges had stood were marked by piles of the bleached bones of existing species of antelope, deer, rabbits, etc. None of these bones were petrified like those at Fossil Lake.

Near the site of each lodge stood a large mortar, made of volcanic rock, with a pestle lying in it.

They had probably been used by the squaws for grinding up acorns and other materials for bread-making. Doubtless a storm of sand had forced the villagers to flee for their lives without giving them time to save even these valuable mortars.

I found a spring of cold water which had built up a mound of white sand, and from the side of a sandhill I pulled out the back part of a human skull. I could not tell how large the village had been, as it extended into the sandhill.

I soon found where the ancient arrow-maker had had his shop by the great quantities of cast-off obsidian chips that covered the ground, as well as by the broken and perfect arrow-heads and spear-points, beautifully polished and finished, and the knives, drills, and the like that lay about. I did not find a vestige of anything made of iron.

Having secured a number of the obsidian points, which I afterwards sent to Cope, I started for camp; but I had delayed too long, and night overtook me before I reached home. My pony and I came near being lost in the desert. I gave him the lines, but I was much worried at not seeing the welcome glow of the camp fire, when I had thought that I must be near my tent. Finally I shouted, and at last heard a faint answer. But even then, owing to my deaf ear, I could not locate the camp, and had to wait until George came up and piloted me in.

Now without doubt the arrow-heads and spear-points mingled with the bones at Fossil Lake are of the same manufacture as those which I found at this Indian village, although the latter are not so much weathered, having evidently been recently covered with sand. I conclude, therefore, that the implements mingled with the bones are no older than the village, perhaps a hundred years old. They were probably shot by the Indians of the village at the wild animals which doubtless came in great numbers to the lake to drink. Then some powerful wind, like that which covered the village, drifted away the sand that lay over the fossil bones, and the flints, being too heavy to be carried away with the sand, dropped down and mingled with the bones. This seems to me the only possible explanation. And I am glad to say that so high an authority as Professor J. C. Merriam of the University of California, after the most careful study and explorations, agrees with me in this. He has recently been over the Fossil Lake region, and he assures me that it is a mistake to suppose that the human implements found there were contemporary with the extinct animals of the Equus Beds.

Whenever George and I had collected a load of fossils, we took them in to Button's ranch. One day we were late in starting, and realized that we should have to hurry to reach the ranch before dark.

As so often happens, this was the very occasion upon which we were fated to be delayed.

At a certain place on our route, we had to pass some mud springs, circular wells filled to the brim with thick, yellowish mud of the consistency of mortar. In wet weather they continually boiled up without overflowing, but to-day they were covered with a hard coating of dry mud, cracked deeply in all directions.

I called to George, who was driving the pack horse, to watch him and see that he did not jump into the spring that we were just passing; but the words were hardly out of my mouth when the miserable wretch made a running jump, and landing in the middle of the crust, broke through and went down into the thick, nasty mud. As he was going down, he seemed to realize what he had done, and managed to get his front feet over the rim of solid earth. And there he hung, the broad pack—we had brought along our tent and blankets—helping to buoy him up.

We sprang from our horses, and made a rush to save our precious fossils, beside which everything else, including the mischievous pony, was of no account. We had to cut the ropes that bound the fossils and camp outfit to the animal, and when we had them safe on solid ground, tie a rope around his neck and pull him out. Of course he was thoroughly frightened, and did everything in his power to help us. Such a looking horse you never saw as he was when we got him out. His whole body was covered with a coat of sticky, yellow mud, which we could not scrape off. We had to take him into a creek and give him such a scrubbing as, I think, no member of the genus *Equus* ever had before or since.

All this took time, and it was late at night before we reached the ranch. It was our habit, when we got to the cabin and felt that it would be too much trouble to open our pack and get out our own supplies, to help ourselves from Mr. Button's store. So, after we had put the horses in the barn and given them a liberal feed of oats and plenty of hay, we went into the larder to get something for our own supper, for by that time we were pretty hungry.

After supper I lay down on the absent lord's blankets, and was smoking the pipe of peace, when a knock was heard at the door. It surprised me, as it was the custom of the country to walk in without the formality of knocking. I shouted, "Come in!" and a short, heavy-set man entered. He said that he had been overtaken by night, and as both he and his team were in need of food, rest, and shelter, he wanted to know whether we would take him in.

"Why, certainly," I answered. I have noticed that most men are liberal with other men's property. "I don't own the ranch, but we have just put our horses in the barn, where there is plenty of hay and oats, and there is plenty of food here. George will show you the way to the barn and help you unhitch, and I will have supper ready when you return."

He thanked me, and while they were putting up the team, I got a hot supper with materials from Mr. Button's larder. This meal was greatly relished by our midnight guest.

I returned to the bed and my pipe, and was entering into a lively conversation with the stranger, when the thought suddenly flashed into my head, What if this man owns the ranch? I sprang from the bed on the instant, and fired pointblank the question, "Do you know Lee Button?"

"Yes, I've seen him," was the answer.

"That's your name, isn't it?" I asked.

"Yes," said the stranger, and I felt so cheap that I would have sold out for nothing. But this was Mr. Button's chance to show what sort of a man he was, and when I apologized for the freedom with which we had made ourselves at home in his house and used his goods, he told me that we had done exactly right, and that he would have felt hurt if we had acted otherwise.

He became a true friend and helper, and his log cabin proved a valuable place of shelter for my party during some of the cold October nights. If these lines should ever reach his eyes, they carry to him my cordial thanks for his hospitality.

CHAPTER VII

EXPEDITION TO THE JOHN DAY RIVER IN 1878

URING the winter 1877-'78 I camped on Pine Creek, Washington, exploring the swamps in the neighborhood and fighting against water to secure specimens. We had dug a large shaft down to the bed of gravel, twelve feet below the surface, in which bones were to be found, but every morning we found that the hole had filled with mud and water over night, and we had to spend hours bailing it out. When we finally got it clear again, we had little time or strength left for securing fossils. This performance had to be repeated day after day, and of course the farther we excavated, the more water there was to be bailed out. I don't think that we were dry a single day that winter. But luckily the water was warm, and we did not suffer from colds.

On the twenty-third of April I started with a team and wagon from Fort Walla Walla, accompanied by my two assistants, Joe Huff and

"Jake" Wortman, the latter at that time an intelligent young man from Oregon, who had been introduced to me the winter before by my brother, Surgeon George M. Sternberg, at that time post surgeon of Fort Walla Walla. During the past six months Wortman had been my guest at my camp on Pine Creek. Afterwards he became known to science as Dr. J. L. Wortman.

We skirted the Blue Mountains in a southwesterly direction, traveling through the beautiful wheat-fields of that fertile region; and striking south at Cayuse Station on the Umatilla Reserve, we climbed the long slopes of the mountains and plunged down into the Grande Rounde, once the bed of an ancient lake, but now a lovely valley nestling among the hills. From this point we drove south to Baker City, and leaving behind us the jagged peaks of the Powder River Mountains, struck the John Day River at Canyon City.

On the second of May we camped on the other side of the mountains in a large meadow. The boys went hunting and got a deer. On the third, our road led us again through rugged mountains, covered in places with ice, and we had to cut foot-holds for our horses, as they were smooth-shod. We passed through a large mining gulch, where men were at work placer-digging for gold. The whole surface of the country had been dug over, and was disfigured with holes and ditches and heaps of earth.

On the fifth of May, after passing through Canyon City, we started for the John Day Basin. It snowed nearly all day. On the road we met a man who told us of a rich fossil leaf locality, on the Van Horn ranch; and after a sixteen-mile drive we found the place and secured some very fine specimens. The leaf impressions were found in a soft, shaly clay-stone, and were very abundant, representing well-preserved Tertiary flora. That night we feasted on a large salmon trout which I caught in an irrigation ditch.

On the sixth (I am following my notebook) we worked all day. I collected two hundred specimens, and Mr. Wortman eighty-five. They were all very fine, and represented the oak, the maple, and other species. I secured some fish vertebræ also. This is another case in which I lost credit for early discoveries. I was told by Professor Cope, a few years before his death, that these specimens had never been examined.

In this same locality there is a bed of rock so light that it floats. I threw a large mass of it at some object in the water, and was amazed to see it float off down the stream. It was the first time that I had ever seen a rock lighter than water.

On the seventh of May, after a journey of fifteen

days from Walla Walla, we reached Dayville, a mile below the crossing of the South Fork of the John Day River. One of the first men I met was a certain Bill Day, whom I soon after hired as assistant. He had for years been making collections of the fossil vertebrates here, usually sending them to Professor Marsh. I was able to secure a large and fine collection from him and another mountain man, a Mr. Warfield, who had also spent much time collecting fossils. Both men had been employed by Professor Marsh during his expedition in this region, and were very careful workmen.

We camped on Cottonwood Creek and prepared to pack into the Basin, or Cove as it has been called. For a hundred and fifty miles of its course, the John Day flows east, skirting the Blue Mountains, but here at Cottonwood or Dayville, it has turned north and cut a great canyon, four thousand feet deep, through the heart of the mountains, the so-called Grande Coulée, since known as the Picture Gorge. At the foot of this canyon, the mountains swing away from the river in a great horseshoe bend, closing in upon it again several miles below. This amphitheater, three miles wide and thirteen long, is a scene of surprising beauty. The brilliantly colored clays and volcanic ash-beds of the Miocene of the John Day horizon paint the landscape with green and yellow and orange and other glowing shades, while in the background, towering upward for two thousand feet, rise rows upon rows of mighty basaltic columns, eight-sided prisms, each row standing a little back of the one just below, and the last crowned with evergreen forests of pine and fir and spruce. But no pen can picture the glorious panorama.

Ever since Cretaceous times, when a quiet inland sea laid down the thousand feet of Kansas chalk, here in the John Day region vulcanism has held sway; almost until to-day. Indeed I have often seen the summit of old Mount Hood wreathed with menacing clouds of smoke, as if she were preparing to pour forth again her floods of molten lava and devastate the region.

When volcanic action first began, great masses of ashes must have been thrown out over the country, settling in the lakes and covering the remains of animals which had been accumulating there for ages. Then floods of lava, one after another, poured out over the forests, until they lay buried beneath two thousand feet of volcanic rock. Where did this immense mass of molten rock come from, and how? A dike crosses the Basin, and for fifteen miles the basaltic columns lie along its edges like cordwood; so we know that some of the lava at least was squeezed up out of the earth's crust through narrow cracks.

I remember once, as I was standing with Uncle Johnnie Kirk, the hermit of the Cove, in front of his cabin, he pointed to the basaltic cliffs that towered above us, and observed gravely, "All vegetable matter." He had found at the base remains of the forests which the lava had engulfed, and had concluded that the whole mass represented similar remains.

Before moving the outfit into the fossil beds I took my pony and started off to spy out the land. Following a horse trail that led up the gentle slope west of the canyon represented in Dr. Merriam's picture of the Mascall Beds I reached a tableland, which proved to be the divide between Cottonwood and Birch creeks. Here I found that the trail leading down to the mouth of Birch Creek was very steep—one could have greased one's boots and slid the whole distance of several hundred feet. I was afraid to ride down and led my pony, but I soon learned that an Oregon pony has long, well-developed legs and can climb up and down better than I could myself.

When I reached the river at the mouth of the Grande Coulée, I found to my dismay that all the rich-looking green and brown fossil beds were on the other side, where the amphitheater which I have mentioned is cut out of the flank of the mountains. As a boy I had learned to swim dog-fashion, and as the river was not over thirty or forty feet wide, and I was determined, after coming so far, to find some fossils and a good camping ground, I decided to strip, jump out as far as I could, and paddle the rest of the way across.

No sooner thought than done. In I sprang, discovering too late that I had reckoned without my host and that the river, which had been penned in for miles by the walls of the canyon, was here flowing away from its prison with amazing swiftness and power. My weak little body was as helpless as a straw in its grasp: down I went, and striking a boulder at the bottom, was flung up five feet into the air, I took in breath and closed my mouth as I went down again; tossing me hither and thither like a cork, beating me against rocks and hurling me high into the air, the river bore me swiftly on, until at last, thank God! it tired of its toy, and threw me to one side into deep water, under a willow whose welcoming branches I eagerly clasped. There I hung until I had regained my strength enough to pull myself out.

But the fossil vertebrates of the John Day beds were still across the river and the questions which I had crossed the mountain and risked my life to answer were still waiting for replies. Unwilling to return home beaten I walked up and down the river shore, and was delighted to find an old boat caught in a pile of driftwood. I dug it out with my bare hands, only to find that its seams had parted and that its bottom was as full of holes as a sieve. Not dismayed, I found a bed of sticky clay with which I calked my ship, and venturing again into the flood, managed to get to the other shore before the boat sank.

I found a place to camp lower down, at the mouth of a canyon which opened out into the level country, and on a little creek that ran in front of Uncle Johnnie's cabin. I was very well pleased with my explorations in the fossil beds also, for I found the skull of an Oreodon, a hog-like creature which, judging from the abundance of skulls and skeletons, must have lived in droves during the time when this rock was being deposited in the lakes of this region. These animals were herbivorous in habit. Uncle Johnnie always referred to them as bears. He often brought a skull into camp with the remark, "Here's another bar's head. I've killed hundreds of 'em in ole Virginia."

I returned to camp much elated, and was planning to pack the outfit into the Basin the next day, when to my disgust Joe Huff, who owned the horses, refused to pack them, as he did not want to run the risk of injuring them. It was useless to tell him that he had been hired to do what I wanted, etc.; he was not to be moved. So I paid him off, and saw him start for his home near Moscow, Idaho, riding bareback. I felt sorry for him, but he had a stubborn fit on, and there was no doing anything with him. After I had hired Bill Day, he wanted me to overlook the past and re-employ him, but it was too late then.

I suppose Bill Day must have weighed about a hundred and eighty pounds, but he was an expert hunter and a keen observer. He owned a herd of ponies and furnished me with all that I wanted, and as he knew every inch of the fossil beds and all the best camping grounds, his services were invaluable. He kept our larder supplied with venison, also. I think my success in that region was largely due to his assistance. I was also indebted to a Mr. Mascall, a man who lived on the second bottom of the river. He had an extra log cabin behind the one he lived in, and he let us use it as a storeroom for our extra supplies of food and for our fossils, when we began to secure them.

This Mr. Mascall had a wife and daughter, and when we came in from the fossil beds, after several weeks of camping out, it seemed almost like coming home to be able to put our feet under a table, eat off stone dishes, and drink our coffee out of a china cup, and to sleep on a feather bed instead of a hard mattress and roll of blankets. Then Mr. Mascall was a good gardener, and always had fresh vegetables, a most enjoyable change from hot bread,

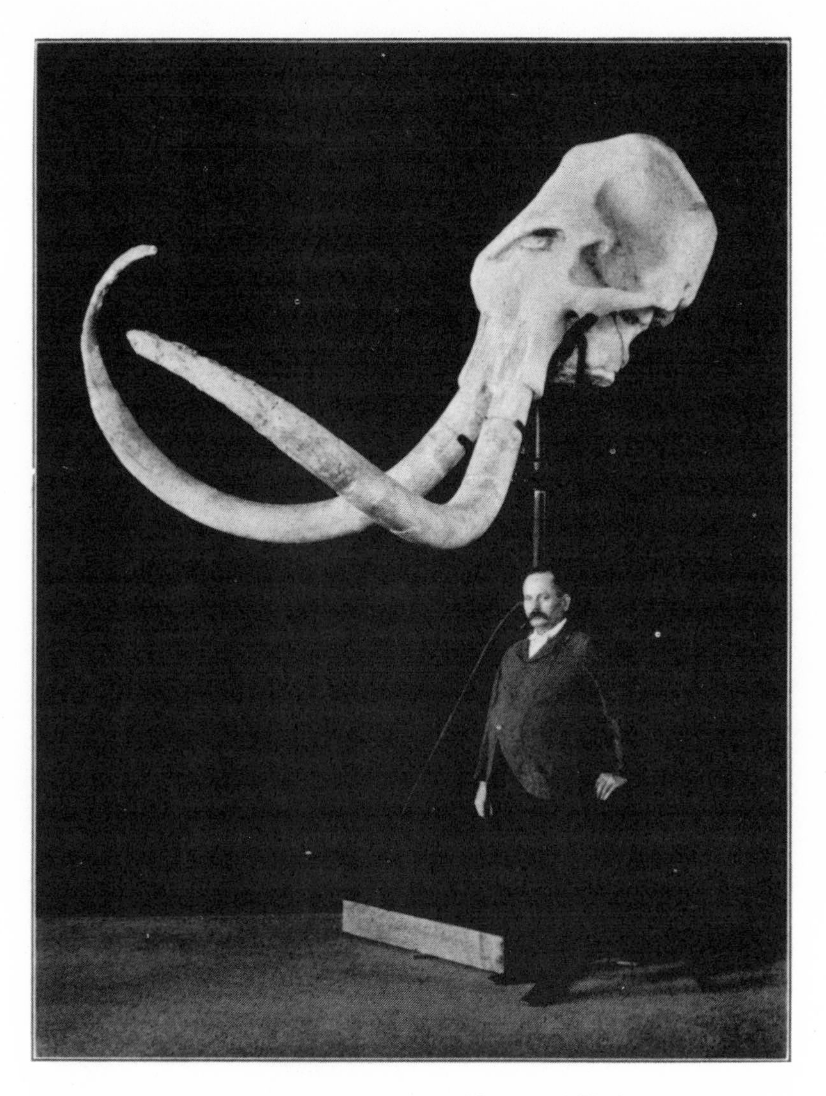

FIG. 26.—SKULL AND TUSKS OF IMPERIAL MAMMOTH, *Elephas imperator.*
In American Museum of Natural History.

Fig. 27.—Fossil-bearing Cliffs. (After Merriam.) Upper John Day exposure.

Fig. 28.—Fossil-bearing Cliffs. (After Merriam.) Middle John Day exposure.

bacon, and coffee. I shall not soon forget his hospitality.

When all was ready, we were taken across the river in Mr. Mascall's boat, swimming our horses. Then the packs were adjusted, and the wearisome climb up the face of the mountains began. It usually took us half a day to reach the summit. Then we climbed down steep slopes and over spurs of the hills, until we reached Uncle Johnnie Kirk's hospitable cabin, a 12 x 14 structure of rough logs with a shake roof. He kept bachelor's hall and lived all alone, except when some cowman or fossil hunter came along. We pitched our tent near his house.

Not far away there was a tract of bad lands, called the Cone, the largest in the John Day Basin, covering, I should judge, a section of land. It was cut into the usual fantastic forms, peaks, ridges, and battlements, and slender spires sometimes a hundred feet high, and as thickly clustered as those of some old Gothic cathedral. Their summits were crowned with hard concretions, which protected their almost perpendicular sides from destruction by the elements.

The drainage canals spread out through this territory like the ribs of a fan, converging at the entrance, and woe to the man who chanced to be caught in one of them during a rain, for the steep slopes shot the water down into them with such amazing rapidity that before he could turn around he would be engulfed in fathoms of water. We always climbed up to some high point the minute we heard the rain strike the rocks above us, and waited until the storm was over and the water had run out. A ditch containing twenty feet, sometimes, of water would dry up as soon as it stopped raining, so steep was the slope of its bed.

I was continually impressed in this region by the power of running water. Not only is this manifested in the mighty canyons which have been carved out during the course of ages from the solid rock, but I stood transfixed with astonishment once, at the mouth of the little creek in front of Uncle Johnnie's cabin, on finding it dammed by a mass of basaltic rock, weighing at least twenty tons, which had been brought from its native hills, three miles away, by a flood of water, and left stranded here. All the side canyons that empty into the John Day River have dumped their loads of boulders there, in some places damming the stream or creating a series of rapids.

I soon found that all the ground in the fossil beds which was easy to get at had been gone over. Here and there we would run across a pile of broken bones and a hole from which a skull had been taken. When I asked Bill what he had meant by leaving the bones of the skeleton behind, he answered, " We were only looking for heads, though we sometimes saved knucks and jints." This accounts for the scarcity of skeletons among the first collections made. I saw to it that my party should care for every bone discovered.

I realized then, that if we were to make our expedition a success, we should have to climb where no one before us had dared to go. It was a serious matter to scale those almost perpendicular heights; one took one's life in one's hand in attempting it. They were, of course, entirely bare of vegetation, and where the slope was not too steep, they were covered with angular fragments of rock which rolled from under one's feet and were likely to send one flying into the gorge below. But I laid the situation before my two men, explaining to them that unless they were willing to face the danger, we should have to give up the expedition, as we had explored the safe ground without results; and they courageously agreed to follow where I led.

So every morning we started out for a day of perilous enterprise, each with a collecting bag over his shoulder and a well-made pick in hand. The latter was used not only for digging out fossils, but was absolutely indispensable as an aid in climbing, and as an anchor in case we began to slip. We were never sure when we left camp in the morning that we should all meet there at night, since a single misstep on those cliffs would mean death or worse than death on the pitiless rocks below; but every day we gained confidence and grew more skilful in the use of our picks.

Far above the pick-marks of the fossil hunters who had preceded us, far above the signs of the mountain sheep that inhabited these wilds, we made our way, cutting niches for our feet as high above us as we could reach, and drawing ourselves up with bodies pressed to the rock. At each niche we rested, and scanned the face of the cliff for the point of a tooth or the end of a bone, or for one of those concretions, among the thousands that everywhere topped the pinnacles or projected from the rocky slopes, whose skull-shaped form revealed the treasure that was hidden away within. When a fossil was found we first cut out of the face of the cliff a place large enough to stand upon, and then carved out the specimen.

I could tell of a hundred narrow escapes from death. One day I was standing on a couple of oblong concretions, about a foot in length, with a chasm, fifty feet deep and three or four feet wide, immediately in front of me. After I had searched carefully the surface of all the rocks in sight, I started to jump over to a narrow ledge on the other side of the gorge. Suddenly both concretions flew from under my feet, and I was plunging head downward into the gorge when by a violent struggle in mid-air I managed to throw my elbows on the ledge; and I hung there until I could find a foot-hold and pull myself out onto solid rock.

Another time I was climbing a steep slope which was capped by a perpendicular ledge. I thought, however, that I could climb over it to the top of a ridge that ran back into the hills, where I could find a way down. For understand, we could never go back the way we had come, as we could not relax our muscles sufficiently to enable us to find with the tips of our toes the niches by which we had climbed up. So we had to be sure that we could get to the top and find a way down from there. On this occasion I was so busy searching the face of the rock for fossils that I worked for hours, climbing up niche after niche, without noticing very much where I was going, until chancing to look upward, I discovered that an escarpment of the top ledge leaned over the slope that I was scaling, rendering it impossible for me to reach the top. I fully expected that I should have to cut out a place to sit in and wait until the boys missed me and looked for me. They could then reach the top of the ledge by some other way, and lower a rope to me. But I was delighted to find at last a perpendicular seam in the rocky ledge, which proved wide enough to admit my body. So I climbed to the top as a man climbs a narrow well, with my back braced against one side and my feet planted against the other.

But such experiences as these, instead of making us timid, only spurred us on to more dangerous attempts. To show how reckless we became, I remember that once Bill found a skull in a perpendicular cliff of solidified volcanic mud, the termination of a ridge that ran far back into the hills. The skull was located about twenty feet up the face of the cliff, and too far below the surface of the ridge to be reached from above; so that there was no way to get at it but by scaling the cliff. I cut niches on one side, and Bill on the other, and we climbed up until we could reach the specimen with our picks, clinging to a niche with one hand and wielding the pick with the other. I worked with my right hand and Bill with his left.

The rock was very hard, and it took a long while to hew out the specimen. While we were at work, we heard a mountain sheep bleating for her young. By reaching up we could get our hands over the edge of the cliff, and pull ourselves up so that we could just peek over. Sure enough, the sheep was coming down the ridge toward us in great excitement, rending the air with calls for her lamb. I began to imitate the bleat of her offspring, and she increased her speed toward us with every sign of relief.

"What if she should butt us off?" I said to Bill, and the position we were in, clinging to the face of the rock with our toes and fingers, made the idea so inexpressibly funny that he began to laugh, louder and louder the more I tried to hush him up. When I had led the sheep up to within ten feet of us, she concluded that we were not her lost lamb, and turning like a flash, started on a run for the mountains a mile away. Out of a side canyon came the lamb, and fell in behind its mother; and we could see the dirt flying out behind them until they appeared to be about the size of a rabbit and a ground squirrel.

One day Bill and I were out together in the beds, and when we got back to dinner, Jake did not show up. We were not much concerned about him, as we concluded that he had found a specimen and was digging it out; but when we came in at night and there was still no Jake, we made up our minds that he had either fallen and killed himself or that he was lying in some gulch with a broken limb. In great anxiety we started out into the Bad Lands to find him.

It was a dangerous enough expedition in the daytime, but doubly so at night, and we risked our lives many times; but we did not give up until we had made the desolate region ring with our calls. At last, about midnight, with fear and sorrow in our heart, we returned to camp. By the moonlight I saw what appeared to be a human form in Jake's bed. I rushed to it and threw off the blankets, and there, sleeping peacefully, lay Jake. We had a great mind to take him out into the Bad Lands and pitch him off into a canyon. It seems that he had been to the mountains, three miles away, where a small exposure of the John Day beds could be seen from camp; and when he returned and we were not in, he had not worried about us, but had eaten his supper and gone to bed, while we were making ourselves hoarse shouting for him. This incident illustrates a peculiarity of youth—its thoughtlessness as to the anxiety which it may be causing its elders.

Among the fossil remains which we secured in these John Day beds, were the limbs of a huge *Elotherium humerosum*, so named by Cope on account of the great process on the humerus. We found the specimen in Haystack Valley, lying on its side, with its toes sticking out of the face of a slope. There were thousands of feet of volcanic rock above it. Following in with pick and shovel, we cleaned up the floor, to find, when we reached the center of the humeri and femora, that they had been cut through as smoothly as if it had been done with a diamond saw. I knew, of course, that there had been a fault here, and that the earth in slipping down had severed the bones. The question that interested me was which side had gone down and how far. If the side toward the open valley, then the rest of the skeleton must have been destroyed by the wash, as the slope above the bones lay at an angle of 45 degrees to the floor on which they lay. If, on the other hand, the mountain side had gone down, and the slip had not been too great, I should be able to find the rest of the bones. Inspired by this hope, we put in several days of hard work, and were delighted to find the severed bones three feet below the original level.

What a shaking and trembling of the earth's crust there must have been, when miles of the mountain mass slipped down three feet toward the center of the earth! No wonder that when a similar fault occurred at San Francisco, the puny works of man fell in ruins. The bones of this *Elotherium* are now on exhibition in the American Museum, which purchased the Cope collection, including the material that I secured through eight seasons in the field in charge of his expedition.

I had found in the Cottonwood beds that lie on top of the John Day Miocene the cannon-bone, or long cylindrical foot bone, of a large camel. As I closely studied this bone, which is composed of opposite halves, separated by a thin septum of bone in the center, with a medullary canal on each side, the conviction came to me that the two halves had once been distinct, like the metacarpals and metatarsals of the pig. With this idea in mind, I was constantly looking for a camel in the older beds, and I cannot express my delight when one day, as I was exploring the John Day beds, I came across a skeleton which had been weathered out and lay in bold relief on the face of a slope. I knew before I picked up the cannon-bone that my belief was verified, and when I took up the two bones separately, the fact was proved beyond a doubt that in this ancestor of the living form the metacarpals of the fore foot and the metatarsals of the hind foot were respectively distinct. As the species represented by this specimen was new to science, Professor Cope named it in my honor *Paratylopus sternbergi.* A skull of this species was afterwards found by Dr. Wortman, and both specimens are now on exhibition in the American Museum.

I arrived at this conclusion with regard to the cannon-bone of the ancient camel as Darwin, Marsh, and Huxley arrived at the conclusion that the ancient horse had three toes. They recognized that the splint bones of the horse represented the side toes of rhinoceroses, one on each side of the middle metacarpals and metatarsals respectively, and they decided that they were the remnants of side toes in the ancestor of the horse. And later we also found a three-toed horse,

I secured also in these beds the skull of a peccary and an oreodont, both new, and used as the types of Cope's description, and a couple of carnivores; one, called by Cope *Archælurus debilis*, about the size of the American panther, the other a dog about the size of a coyote. Cope gave the name *Enhydrocyon stenocephalus* to this genus and species. A splendid skull of the rhinoceros *Diceratherium nanum* Marsh, was another of my discoveries here. All the specimens, with the skull of a rodent from the same beds, are now on exhibition in the American Museum.

Of course these are but a few of the many specimens secured in these beds; hundreds are stored away in the drawers and trays of the Museum. I was told that it would cost twenty-five dollars to get a typewritten copy of the list of John Day fossils in the Museum. In that list are many specimens which my party secured or which I purchased from Warfield and Day. Professor Cope once wrote me that my collection there represented about fifty species of extinct mammals.

One day in July I left Jake Wortman in the field and started for Dayville, leading a pack pony. I intended to stay all night with Mr. Mascall, leave my load of fossils, and take back a load of provisions. Bill Day had lost one of the horses, and as a large band of Umatilla Indians was encamped on Fox Prairie at the summit of the mountains, about six miles east of our camp in the Cove, he had gone off in that direction to look for it.

When I reached the high mountain above Dayville, I could look down into the narrow valley of the John Day. Although it was noon, there was no smoke rising from the chimneys of the houses. The fields of wheat were ripe for the cradle—they had no machines in that region, and not only cradled their grain, but threshed it with horses, who tramped it out—but no one was working in them, and there was no stock in the pastures. What could it mean? I asked myself; and as I followed the long trail down to the river, my heart was full of fearful forebodings. Had a pestilence killed all these people whom I knew so well? Or had they all fled, with their horses and cattle, from Indians on the warpath?

Without expecting to hear a response, I called, when I reached the river, for Mr. Mascall to come over with his boat and take me across. To my delight, I saw him come out of his house and take the trail down to the boat through the woods that covered the first river bottom. All the while that he was unlocking the boat and rowing across, I kept shouting, "What's the trouble? Where are all the people?" But not until I had got aboard with my pack and saddle, and we had started back, would he answer the questions which I had been asking myself ever since I left the top of the mountain.

It seems that three hundred Bannocks, or Snakes, under their chosen leader, Egan, had left the Malheur Agency, several hundred miles south, and after stealing six thousand horses, mainly from the French brothers' ranch, were now on their way north to join Homely, the chief of the Umatillas, at Fox Prairie. General Howard, who was in hot pursuit, had sent a courier ahead of his command to the settlers in the John Day valley, advising them to gather at some central locality, build a stockade, and take their women and children into it for protection from the treacherous redskins. Everyone in the valley, except Mr. Mascall and an old man who kept the mail station on Cottonwood Creek, a mile to the south, had taken this advice and gone to Spanish Gulch, a mining town on top of the mountains about ten miles southwest.

Near sundown Bill Day came in, having heard the news at the Indian camp. He instantly insisted that we leave everything and go to Spanish Gulch. It was foolish, he said, to risk our lives going back to warn Jake. On the long trail up the mountain we should be in full sight of the South Fork, down which the Indians were expected to come, and it would take us half a day to climb those four thousand feet and hide ourselves in the canyons on the other side. I refused, however, to be moved by his arguments. I told him that I meant to go back, and that he was to go with me. We could not leave Jake there in camp, entirely unconscious of the fate that might be approaching him. He knew nothing of the proximity of hostile Indians, and it was our duty to warn him.

"Well," Bill said, "I am going to look out for number one. I have not lost any Indians. If you have, go and hunt trouble. Let Jake look out for himself."

All my shells, perhaps three hundred, were empty, but I had plenty of powder and lead, and the best long-range rifle I had ever owned, a heavy Sharp's weighing fourteen pounds, and shooting a hundred and twenty grains of lead and seventy grains of powder. I set to work cleaning and oiling it; and then spent the whole night in front of the fireplace, melting lead, casting bullets, and loading shells. Bill also stayed awake, and with his needle-gun kept guard at a porthole which commanded a good view of the open ground around the house.

The next morning I started alone on my pony to follow the trail to the Cove, where Jake, unconscious of danger, was at work in the fossil beds. It seemed an interminable journey, and I thought that there was an ambuscade behind every bush and pile of rocks that guarded the road. But, greatly relieved,

I got out of sight at last in the deep canyons on the other side, and soon saw Jake's pony near a fossil bed and found Jake himself deeply interested in a splendid discovery he had made.

When I told him the news, he wanted to drop everything until the war was over, and fly for safety to the stockade. But no; my tent, with many fine fossils in it, was in an open valley in plain sight for miles, and would quickly attract any marauding hostile, who might set fire to it and destroy the work of months. I insisted, therefore, upon caching, the Pacific coast term for hiding, everything. So we took down the tent, and putting it, with the fossils and all the rest of the outfit, into a secret place, we covered them with a big brush pile. Then I was ready to fly as fast as our ponies could carry us.

When we reached the river, Bill was still with Mr. Mascall, and brought over the boat. Then both men insisted that we go without further delay to the Gulch, as we had risked our lives long enough. But there was a large collection of valuable fossils in the log house behind Mr. Mascall's cabin, and as the specimens were wrapped in burlap, they would be destroyed if the Indians burned down the house, which they would be sure to do if they came. I had no boxes, but I had a quantity of new lumber, which we had secured from a mill in the vicinity; so, refusing to be moved, I took off my coat and went to work sawing up the lumber and making boxes. The other men never let their guns leave their hands, and kept guard all night, expecting every moment to hear the whoop of the Indians.

By daylight I had every fossil neatly packed, each in a little box, and then we all took hold, and carrying the boxes down to the first river bottom, hid them under a great grapevine, which completely covered them. After throwing dead leaves over our trail, I was satisfied that we had done all that we could, and as we could not induce Mascall to abandon his property, we left him and went over to the Gulch. We found nearly all the settlers keeping house inside the stockade, which was built of pine logs and covered enough ground to hold their teams, wagons, and cattle, as well as themselves.

As I realized that it would be impossible for us to do any work in the John Day beds, fearing every moment to be surprised by Indians, I concluded that this would be a good time to go to the Dalles and try to find out what had become of the collection of Fossil Lake material which had been sent off the year before, and had been lost somewhere. I had a receipt for the specimens from a Mr. French, who was, I supposed, the agent for the Oregon Steam Navigation Company. His letterhead read "Forwarding Agent for the O. S. N. Co.," but I had repeatedly written to the agent at the Dalles, and had received no answer, while Cope, from his end of the line at Philadelphia, had sent tracers out over every route he could think of, trying to locate the fossils.

A Mr. Wood, the owner of a large herd of horses, was driving the herd to a point near the Dalles for protection from the Indians, and I joined his party. But the several hundred horses raised such a volume of dust that, after a few days of suffocation, I concluded that I might as well lose my scalp as be choked to death, and leaving the herd, went on alone. All along the way, men, women, and children were fleeing for safety to the Dalles, and dozens of homes and ranches were being deserted just at the time when the people should have been saving their grain. I never in my life saw so much excitement and fear. As many white men were fleeing for their lives as there were Indians on the warpath, and every man of them was blaming General Howard for not having exterminated the hostiles before they started.

I met the man who had hauled my Fossil Lake collection in to the Dalles, and for the first time learned the truth about them. It seems that they had never been shipped. Mr. French simply had a warehouse, and forwarded goods by the Steam Navigation Company, and mine had been covered up in the warehouse and entirely forgotten. I was in splendid spirits when I knew that they were safe.

Having rescued this valuable material from the warehouse, I returned to the Gulch without seeing an Indian, to find the people still in a state of great excitement. General Howard had sent word that the men could put themselves under the leadership of Colonel Bernard, each citizen furnishing his own mount and arms, but receiving his rations from the Government. I tried to raise a company of men to accept this offer, but not a man cared to go. At last, heartily tired of staying in camp, I asked for a volunteer to go with me to the John Day valley to find out how Mr. Mascall and the old man at the stage station were getting on. No one would go at first, but later Mr. Leander Davis, who was for many years a fossil hunter for Professor Marsh, agreed to go with me; and packing a horse with blankets and supplies, we started.

We were relieved to find both men well, and no sign of Indians. Continuing our journey east, we crossed the south fork of the John Day, and all doubts as to the movements of the Indians were removed. For a wide trail, cut deeply into the dry soil by six thousand horses and the three hundred Indians who were driving them north, led down the slope and followed up the main fork on the Canyon City road.

As we sat on our horses, looking south along the heavy trail, we saw some half-dozen horsemen coming toward us. We knew that they must have seen us, and concluded to stay where we were until we could make them out. Before long we saw the glitter of sabers and the flash of gold buttons, and soon General Howard and his staff rode up at a gallop. I recognized him by his brigadier general straps and by his empty sleeve. He had lost an arm fighting to preserve the Union.

We saluted, and he asked me whether we had seen his pack train. When I answered no, he asked me if we knew where he could find some bacon, as he and his staff, as well as the troops behind them, had been living for three days on fresh beef without any salt. I told him of a smokehouse across the bridge, and he sent his scout to examine it. The man returned shortly with the report that not only was the smokehouse full of bacon, but that the table in the dwelling house was set for a meal, with cold coffee in the cups, bread, cold bacon, and potatoes, all ready to eat. The people had evidently just sat down to dinner when someone had rushed in with the news that the Indians were coming, and they had all thrown back their chairs and fled for their lives.

While the General and his staff sat down to a hearty meal, Leander and I continued to follow the trail. At one place, where a farmer made cheese, we found that a number of large cheeses had been taken out into the road and rolled along for some distance with a stick. We followed up the trail which they had made in the deep dust, and put one of them on our pack. We went into one of the houses on the road, and found that the Indians had broken up all the furniture, including the sewing-machine, etc. In the front room they had poured out a barrel of molasses, spread over it several sacks of flour, and stuck a little woolly dog in the mixture. The poor little fellow was dead. A little farther on, a sheepman's house had been burned, and near by two thousand sheep had been mutilated and thrown into piles to die. The herders were found scalped a few days later. At one farmhouse a fine brood mare had been killed because she could not keep up with the herd.

Some days later, on the twenty-ninth of July, I believe, there was a total eclipse of the sun. The heavens were like brass, and there was a peculiar condition of the atmosphere such as I have never experienced before or since. A report was spread abroad that the Indians had returned and burned all the farmhouses along the river. I was at the time with Leander Davis, and we rode up to Perkins ranch, where a lot of men had congregated and were taking turns standing guard for fear of the Indians. When we rode up they were standing about, uncertain as to what it all meant. The dogs had gone under the stoop and the chickens to roost. The air was motionless, and an unusual stillness was over everything. The men welcomed us in hushed voices.

I sprang from my horse and asked Perkins whether he had any pieces of broken glass. He said that there were plenty under the west window, and I went and got a supply, followed by all the men, who were greatly relieved by my explanation of the phenomenon. We got a candle and blackened the pieces of glass, and watched the progress of the eclipse through them.

It had a more disquieting effect upon the hostile Indians. It seems that the soldiers had cut them off from crossing the Columbia by capturing all the small boats and patrolling the river night and day; so that with Howard's troops on the trail behind them, troops from Walla Walla on their flanks, and the river in front, they were in a bad way. Moreover, the French brothers and the governor of Oregon had offered a reward of two thousand dollars for Egan's head.

The Umatilla Indians were accused of pretending to help the whites in the daytime, and really helping the Snakes at night. So the commander sent out a party of soldiers to capture the squaws and little children of Homely and the other chiefs and hold them as hostages for the good behavior of their braves. When the latter asked the commander to release their families, the answer was given that if they would capture Egan and deliver him up to the authorities, they would not only get back their wives and children, but would receive the two-thousand-dollar reward. Otherwise their families would still be held as hostages.

It appeared that Egan had an appointment with Homely at a certain hour. As he rode out from his camp, with a brave behind him, Homely, similarly attended, went out to meet him. When they met between the two camps, they turned at right angles and rode toward the point agreed upon for the powwow. But as they were riding thus, side by side, Homely, with a word to his brave, suddenly raised his rifle and shot Egan, while his brave shot the attending Snake. They then immediately severed the heads of the dead men, and riding back with them to the whites, claimed the reward. About the same time, the eclipse came on, and the poor Snakes, deprived of their leader, thought that the world was coming to an end, and leaving their great herd of stolen horses, fled in small bands toward the Malheur Reservation, and were all eventually captured.

The war thus ended, as soon as I could get things in shape and my party together, I returned to the

Cove, got my outfit and fossils, and moved over into Haystack Valley. I remained there all winter, and the next season secured another large collection. Many of the specimens in it are described by Professor Cope in Vol. III of the "Tertiary Vertebrata." On p. xxvi and the two following pages of the preface, he pays his collectors a high compliment, which I give myself the pleasure of repeating here in his own words: "The same year ['77] I employed Charles H. Sternberg to conduct an exploration of the Cretaceous and Tertiary formations of Kansas. After a successful search, I sent Mr. Sternberg to Oregon. The Tertiary formations explored in 1878 were the John Day and Loup Fork of Oregon. The John Day formation was chiefly examined on the John Day River and the Loup Fork beds at various points in the same region. These yielded about fifty species, many of them represented in an admirable state of preservation."

After mentioning the work of his other explorers, he goes on to say: "Mr. Sternberg's expedition of 1878 was interrupted by the Bannock war, and both himself and Mr. Wortman were compelled to leave their camp and outfit in the field and fly to a place of safety on their horses. It is evident that an enthusiastic devotion to science has actuated these explorers of our western wilderness, financial considerations having been but a secondary inducement. And I wish to remark that the courage and disregard of physical comfort displayed by the gentlemen above referred to are qualities of which their country may be proud, and are worthy of the highest commendation and of imitation in every field."

Before leaving this interesting field, I wish to show my readers Cope's figure of the great sabertoothed tiger, *Pogonodon platycopis* (Fig. 31), which was secured in 1879 by Leander Davis. I do not remember who first discovered the specimen, but for weeks each of us collectors, Wortman, Davis, and I, tried to devise some means of securing it. The skull topped a pinnacle, perhaps thirty or forty feet high, and tapering like the spire of a church. At the top it was only a foot in diameter. We knew that it would not be strong enough to support the weight of a ladder, and it was too steep to scale. Moreover, if we blew it up with powder, the skull, whose rows of teeth seemed to grin at us defiantly, would be shattered to bits.

By whatever method it was secured, it represented a feat of the greatest possible bravery, and Cope did only justice to Leander Davis in publishing his understanding of the manner in which it was done. That description is attached to the skull to-day, and thousands have read of Davis' heroic act in securing

FIG. 29.—FOSSIL-BEARING CLIFFS. (After Merriam.)
Mascall Formation.

FIG. 30.—FOSSIL-BEARING CLIFFS. (After Merriam.)
Clarno Formation.

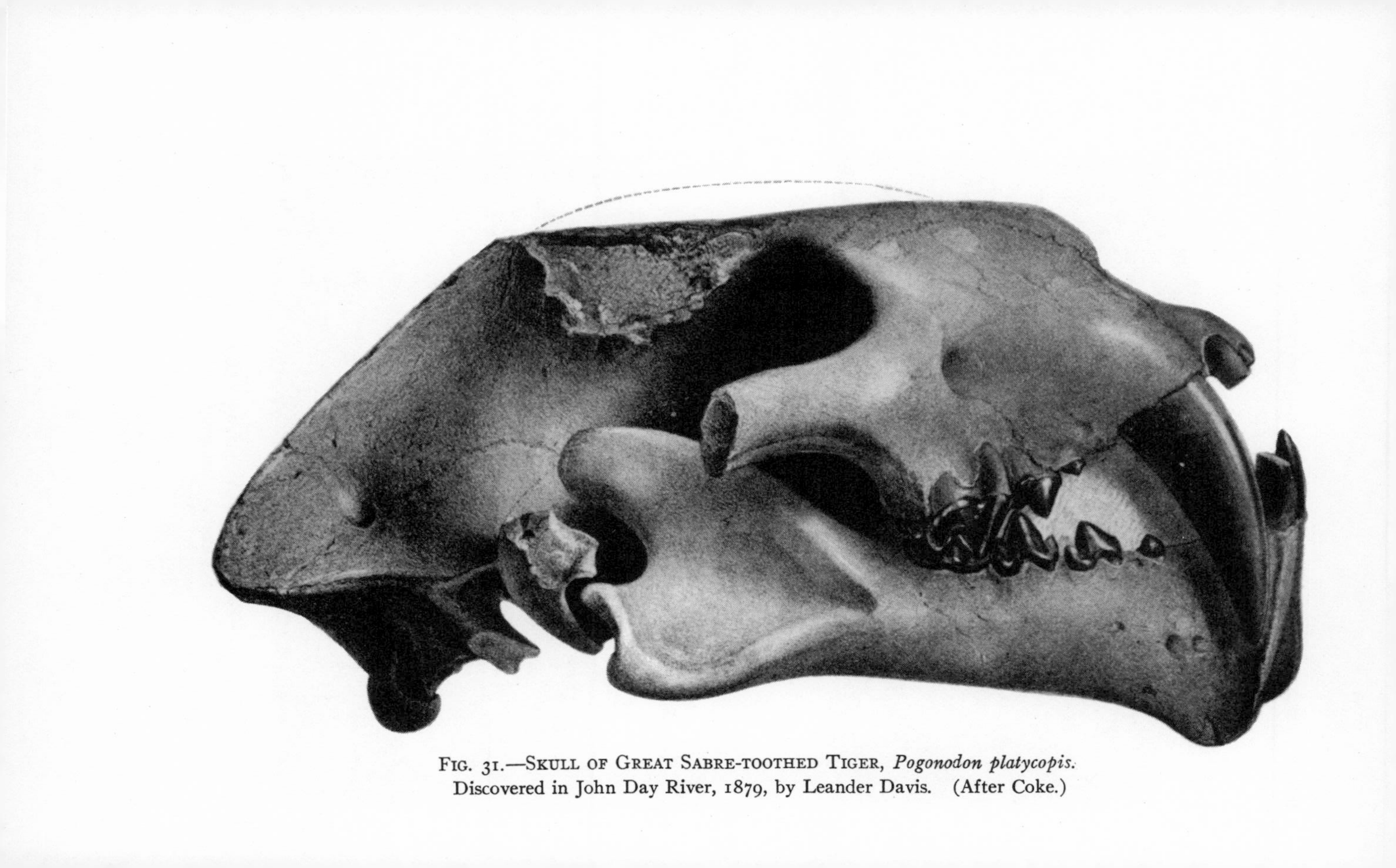

Fig. 31.—Skull of Great Sabre-toothed Tiger, *Pogonodon platycopis.* Discovered in John Day River, 1879, by Leander Davis. (After Coke.)

it for science. Professor Cope says that he cut niches and climbed to the top of the spire. My remembrance, however, is that he threw a rope around the spire and let it settle down to where he thought the rock would be strong enough to support his weight. He then climbed up hand over hand to the loop, stood erect, picked up the skull, and without putting any pressure on the rock, got back to his rope and down to safety below. He then secured the rope by jerking off the top of the pinnacle.

It matters little how he got the skull, but I am ready to testify that it was the bravest undertaking I ever saw accomplished in the John Day beds; and as long as science lasts, this noble specimen of one of the largest tigers that ever lived should be associated with the name of Leander Davis. I am glad that the great dike across the Cove is named after him also.

What is it that urges a man to risk his life in these precipitous fossil beds? I can answer only for myself, but with me there were two motives, the desire to add to human knowledge, which has been the great motive of my life, and the hunting instinct, which is deeply planted in my heart. Not the desire to destroy life, but to see it. The man whose love for wild animals is most deeply developed is not he who ruthlessly takes their lives, but he who follows them with the camera, studies them with loving sympathy, and pictures them in their various haunts. It is thus that I love creatures of other ages, and that I want to become acquainted with them in their natural environments. They are never dead to me; my imagination breathes life into "the valley of dry bones," and not only do the living forms of the animals stand before me, but the countries which they inhabited rise for me through the mists of the ages.

The mind fills with awe as it journeys back to those far-distant lands. Stop, reader, and think! In this John Day region, ten thousand feet, or nearly two miles, of sedimentary and volcanic rock lie above the Niobrara Group of the Cretaceous, from which I dug last summer the beautiful skull of a Kansas mosasaur, *Platecarpus coryphæus,* which lies before me now, its glistening teeth as perfect as in the days when they dripped with the blood of its victims. How many ages were those ten thousand feet in building? How long has it taken the running water, with its tools of sand and gravel, to carve out the Grande Coulée and the river valley, and expose all the various formations, with their records of the life of the past? And yet all this has taken place since my mosasaur, which seems to watch me as I write, fought its last battle and sank to rest beneath the waves of the Cretaceous sea.

CHAPTER VIII

MY FIRST EXPEDITION TO THE PERMIAN OF TEXAS, 1882

Y first expedition to the Permian of Texas was made in 1882, while I was in charge of collecting parties for the Museum of Comparative Zoology of Harvard University.

I left the station at North Cambridge about the fifteenth of December, and reached Dallas on the twenty-first, with the address of A. R. Roessler; but I was told at the post-office that there was no such man and no such address in the city. I had been depending absolutely upon the information which I hoped to receive from this Mr. Roessler, as I myself had no more idea as to the whereabouts of the Permian beds than a new-born child. Dr. Hayden had written me to follow up Red River until I found the red beds, which had colored the whole flood-plain of the valley, and I had seen the red mud at Texarkana as I entered the state; but it would take years to explore the whole valley of that great stream. I felt that I had come upon a wild-goose chase, and I suppose showed my dismay in my face, for the postmaster asked if he could help me. I told him my troubles, and he said that there was a man in town, a Professor W. A. Cummins, who had been Cope's assistant the year before.

Greatly cheered, I went to the man's house post-haste, to be met at the door by his wife, who told me that the Professor was in Austin. Whereupon my spirits dropped below zero again. But if a girl's face is her fortune, so is a man's sometimes, for I gained Mrs. Cummins' sympathy at once. When I told her why I had come to Texas, she answered, "Why, I was with Professor Cummins on his expedition to the Permian beds," and proceeded to give me all the information which I thought necessary.

I learned that they had made their headquarters at Seymour, in Baylor County, between the Brazos and Wichita rivers, and I supposed that anyone in Seymour could tell me the exact localities from which the fossils came. Later I found to my sorrow that this was not the case; and I wasted months of careful exploration over barren beds before I found the horizon that yielded the wonderful batrachians and reptiles of which I had come in search.

Much elated, I took the train for Gordon, a cattlemen's town south of Seymour, and the point nearest to it by rail. I arrived there on Christmas Eve. I was the only passenger to leave the cars and was welcomed by about twenty cowboys, who were just beginning to paint the town red. The leader asked me where I came from, and I answered promptly, "From Boston."

"Where do you want to go?" he asked.

"To the best hotel in town," said I.

"All right!" he said. "We'll take you there." And sure enough, they did. They formed in double file and put me in the middle of their ranks. Then the two men ahead of me laid their Winchesters over my shoulders from in front, and the two men behind crossed these guns with their own, and at the word, "Fire at will!" the whole command opened fire and kept it up all the way to the hotel. There a girl appeared, carrying a lamp with no chimney, and the men, facing the porch, allowed me to go into the waiting room. I turned first, and made a little speech, thanking them for their kind reception and remarking that if I were not so poor, I should stand treat for the whole crowd. This satisfied them, and shouting "All right!" they went off to continue their nonsense until they were all drunk.

I hired the son of the hotel keeper, a Mr. Hamman, put my baggage in his wagon, and started on the journey north to my headquarters at Seymour, which we reached eight days later. Here I got off the track again, for although everyone in town knew Professor Cummins, no one could tell where he had found his fossils. "Over in the brakes," was all the information anyone could give. Finally a man named Turner asked me to come over to his cattle range on the middle fork of the Wichita, as the country was cut up into canyons and ridges and denuded, so that I should be likely to find fossils. He knew of some mastodon bones in the vicinity, he said. So I went with him.

At one place the road led us across the narrows, where there is scarcely room for a wagon road between the brakes of the Brazos and the Big Wichita. Looking south, shallow ravines led to the valley of the Brazos, while to the north were deep gulches and mounds capped with white ledges of gypsum with red beds of clay below. I had reached at last the red beds of Texas.

An interesting phenomenon is to be observed here—the bed of the Big Wichita is one hundred and seventy-five feet lower than that of the Brazos. North of the Brazos, along a line that extends through Baylor County, the country has been lifted up and disturbed by pressure from below, while south of that line, the only disturbance in the strata has been due to erosion. Everywhere in the red beds of the Wichita valley are signs of an elevation of the earth's crust, and for miles down the stream one comes upon miniature mountains with the strata turned up at all angles. The river valley occupies a fault.

Very beautiful indeed was the view when we got in sight of the brakes of the Big Wichita. As far as the eye could see stretched miniature Bad Lands, with rounded knobs, deep canyons, bluffs, and ravines. The prevailing color of the strata was Indian red, but beds of white gypsum and of greenish sandstone relieved the sameness. Sometimes seams of gypsum filled cracks in the strata, forming dikes a few inches in thickness.

Between the hills grew patches of grass, a welcome sight to our horses, for we had passed through a country devoid of vegetation. The fall before, the army worm had eaten the ground clean of everything that was eatable. We pitched our camp near a ditch that had been cut through the sediment which overspread the flood-plain.

The day after pitching camp, I heard George Hamman calling me, and crossing the bridge, saw him beckoning me to follow him. He gathered his pockets full of cobblestones as he went along, and when he reached the edge of the ditch a little way below the crossing, he began to throw the stones at something. I ran up to him, and heard the rattle of snakes, but could not see any until, resting my hand on his shoulder, I lifted myself on my toes and saw, on the other side of the ditch, a cave with a broad floor. Lying singly or knotted together in gorgon spheres, with heads sticking out in all directions, were hundreds of large rattlesnakes, which had come out of the cracks in the earth to bask in the sun on this sheltered floor. They had become terribly irritated by the blows of the stones which Hamman was hurling at them, and were rattling in chorus and striking out in all directions, biting themselves and each other. Suddenly one rattled in the high grass at our very feet, and looking down, we saw a big fellow making ready to strike. As quick as a flash Hamman threw himself over backward, knocking me down, and the instant he touched the ground, turned a complete somerset. While I lay there, overcome with laughter, he turned two more, and finding himself on the road, started for camp on a run. I was too hysterical with laughter to help myself, and lay there, while the snake continued to sound its rattle and dart out its forked tongue, swinging its head back and forth above its coiled body. When George saw my predicament, he was brave enough to come back and pull me out of reach of his lordship's fangs. Then we were mean enough to kill him. He measured five feet in length.

The valley contained thousands of wild turkeys, and it was a fine sight to see them come down in great droves from the hills at night to roost in the trees below. On the level prairie there were many antelope, also; and wild cats and coyotes were seen nearly every day. I remember one day, when crossing a low level prairie covered with bushes a couple of feet in height, seeing at my left a coyote which was running along in a straight line, with its nose pointed toward a certain spot, like a pointer dog after a prairie chicken. My interest was aroused, and to increase my curiosity, I caught sight of a short-tailed cat, the Canadian lynx, crawling along the ground in the same direction. I knew that they were both trailing some prey which each, unknown to the other, had scented, and imagining that it might be a calf, I shouted, as I did not want to see it torn to pieces. This startled the cat, and drove her off at a tangent to her trail. The coyote continued his course, but did not stop, for a Texas cow had run to the point toward which he was traveling, and stood with lowered horns, ready to repel his assault; while her calf sprang up and deliberately proceeded to take advantage of the situation to get his dinner.

In this region, as in the Kansas chalk beds, the question of water gave us a great deal of trouble. All the water in the river is that which goes by the name of alkali in the West, being thoroughly impregnated with salt and other mineral ingredients.

There are, moreover, no wells or springs in the red beds. The surface rock is porous, and the water sinks through it to the compact gray beds below, from which it drains off into the river. These gray beds are some distance below the surface, and so far as I know, have never been reached in digging for water. One is, therefore, forced to depend upon rain water. This is collected either in artificial tanks built by the cattlemen, or in natural tanks, sometimes along the creek beds, but usually in the flood-plain in old creek beds, where the fine red mud has been puddled by cattle, perhaps, or in the olden days, by buffalo. These ponds hold water for years, although often they become very foul from the cattle that frequent and wade into them in summer to get away from the flies.

It is an odd sight to a stranger in the valley of the Big Wichita to see the rain come rushing down the hills. It soon becomes as thick as cream with the fine red clay, and to think of depending upon such water for drinking and cooking purposes is revolting to one who remembers the sparkling springs and clear wells of the East or any mountainous country. During quiet days, when the wind was not blowing, the red mud would settle in the bottom of the tanks, but one had to be careful not to pull out one's pail suddenly or the water would instantly thicken with mud from the bottom.

Nothing would settle this water but boiling it, although it might be cleared a little by the pulp of cactus leaves. I have sometimes gone to the trouble of peeling the broad leaves of the prickly pear and beating them into a mucilaginous pulp to throw into a pail of muddy water. The mud attached itself to this material and sank with it to the bottom; but even then the clarified liquid remaining on top did not make a very tempting drink. I soon got used to the thick red water, however, as had the other inhabitants of the country, and for six seasons drank it thankfully, when I was thirsty. When a man is thirsty, he drinks first and tastes the water afterwards. I once asked an old cowman what he did for drinking water on the range, and he answered, "Wherever and whatever a cow can drink, I can." And cows will take filthy water, if they can get no other.

All that winter I worked in these desolate beds, walking over thousands of acres of denuded rock, searching without success for the fossil fields. The dominant color of these beds is red, but the tints vary so that the eye is dazzled and wearied by the constant change. There are countless concretions too, all of which had to be looked over. If fine specimens had rewarded the labor, all would have been well, but I know of no work more trying than spending day after day in a fruitless search.

At last Hamman, having fattened his horses on two-dollar corn, started a quarrel with me, so that he might have an excuse for deserting me, and drove off with the team, which I had hired for some time longer, leaving me alone, thirty miles from town. Fortunately, however, I found a good, honest Irishman, Pat Whelan by name, who became not only a splendid assistant, but a true friend. Poor fellow! I learned a few years ago that he had frozen to death in Montana.

One warm, sultry day I sent him in to town for provisions. I had no tent at that time, but he left me the wagon sheet, and I had camped on the south side of a large tree, which was so effectually covered with green briers as to be an almost impenetrable defense against the north wind.

I was in the field after Mr. Whelan left me, and noticing the Texas cattle coming from the prairie to the heavy timber, I concluded, although there was not a cloud in sight, that they had scented a norther. Rushing to camp, I began rapidly to make preparations for the storm. First I cut a couple of crotches and sank them well into the ground on the south side of the brier-covered tree. Then I put up a ridgepole and stretched over it the wagon sheet, which I fastened securely to the ground on either side. I also heaped dirt on the edges, to keep out the snow. I thus had a dog tent, opening toward the northern barrier and toward the south.

There was plenty of fallen wood lying about, and I devoted every moment and all my strength to cutting it up and dragging it to the tent. I must have got several cords together before I heard the wind howling in the heavy timber to the north. I piled up this supply of fuel at the opening toward the green brier thicket, and built a big fire at the mouth of the tent.

Soon an awful storm was upon me, all alone, thirty miles from any human habitation. How the wind moaned through the creaking branches! A dense darkness spread like a pall over the heavens, and the shrieks and wails of the tempest echoed through the woods like the cries of lost souls. Then snow and sleet began to fall in fitful gusts, and beat upon the thin canvas that was my only shelter. At such a time a man loses much of his confidence in himself. Pretty small I felt myself when measured with that storm, which bent the great cottonwoods and elms like reeds before it.

After supper, tired out with my unwonted exertions, I fell asleep. Whenever the fire sank down and the cold became severe, I roused myself and piled fresh fuel on the dying embers, and when they blazed up again, dropped off once more. Three days and three nights that norther lasted. I understood then why the people of the Southland speak of them as they do and dread their coming. I never once left my shelter until it cleared.

Poor Pat Whelan! He had lost his horses in the storm, and being sure that I would freeze to death if he could not get back to me, he had spent every hour of daylight looking for them. What he must have suffered in that awful gale, while I was safe and comfortable!

My readers would grow weary if I told the whole story of that winter's search. There were so few results that I became thoroughly disheartened and anxious to give up the fight and go home, where my wife and dear baby were waiting for me. There was further cause for discouragement in the fact that Pat had only agreed to stay with me until spring plowing began, and the time for that was rapidly approaching. But I would not give up. So we worked on down the stream toward the Fort Sill cattle trail, traveling on an average twenty miles a day on foot, with the record " Nothing " in my notebook night after night.

But on the eleventh of February, after forty days of unceasing effort, I discovered below the forks of the Big Wichita a somewhat different horizon from that of the beds over which I had been working so persistently without success. Some of the beds in this region are composed of red clay, with small irregular concretions that are piled in heaps at the base of the hills and roll under one's feet, rendering travel difficult. In other strata are deposits of small nodules, held together by silica. These nodules are of various colors, and where held securely and ground down, make beautiful mosaics. Then there are beds of greenish sandstone, laid down in thin layers; and in these beds, for the first time since I came to Texas, I found the remains of a Permian vertebrate. My notes say: "Although it is not wise to shout before I am out of the woods, yet I feel very much encouraged, and I earnestly hope for the success I have worked for. I have evidently worked too high in the red beds to find fossils."

On the second day in these beds, I found fragments of the great salamander *Eryops*, and on the twenty-second of February, I found the first specimen that I had ever seen of the long-spined reptile, *Dimetredon*. Of this last I got seventy-five pounds of bones and matrix, preserved in iron ore concretions. The teeth are long, recurved, and serrated. I knew little then about these most ancient of all the vertebrates that it has been my fortune to collect, but I shall have more to say about them later. The authorities now place the time when these animals lived twelve million years away. Indeed, "God is not slack as some men count slackness, one day is with the Lord as a thousand years, and a thousand years as one day."

The only way in which we can realize the lapse of millions of years is by a study of the work which nature has accomplished in them, depositing vast strata, lifting them up into mountain ranges, and carving out in them flood-plains and mighty canyons. More interesting still is a study of the countless forms of life which, in ever-varying groups, have each in turn dominated sea and earth and air. First, as here in Texas, the batrachians reigned supreme, a race of creatures which were supplied with both gills and lungs, so that they could live both on land and in water. Then came the reptiles, and later still dawned the Age of Mammals, with man as the crowning work of the Creator's hands.

I was now at last in the fossiliferous beds and secured some fine material. Unfortunately about this time Pat gave notice that he would soon be obliged to leave me. I should then have no team, and to work in these fossil beds without a means of transportation would be as useless as to attempt to dig up a forest with a hoe. I had, however, sent north for an assistant, a Mr. Wright, and after hunting for me a day and a half in the brakes of the Big Wichita, he finally arrived in camp.

On the sixth of March a violent norther struck us. We were better off for protection than we had been, however, as my tent had at last arrived from Kansas; and although only an A-tent, it kept out the storms of sleet and snow that fell for three days. During all that time the cattle remained without food in the dense woods. Such times as this, when we were confined to the close quarters of our tent and could accomplish nothing but keeping ourselves warm, are in my opinion the most uncomfortable which the fossil hunter is called upon to endure.

On the ninth of March, the sun rose bright and clear upon a scene of surprising beauty. Every tree, bush, and blade of grass on the red beds was covered with a milky white ice, whose silvery luster was set with innumerable sparkling gems. It was glorious at sunrise, but as the morning advanced, the snow and ice began to melt, leaving patches of red and white over the Bad Lands, and by noon had entirely disappeared. The hills rapidly dried, as the thick red water sought the drainage canals, and we were soon at work once more.

As a precaution against the very difficulty which I had encountered,—I mean the impossibility of keeping a man and team with me,—I had obtained from the Secretary of War, through the efforts of Professor Alexander Agassiz, a letter of introduction to the commanders of western posts, requesting them to assist me by every means in their power not inconsistent with the public service. With this letter from the Honorable Robert T. Lincoln, a son of our martyred President, I started out on the twelfth of March for Fort Sill, on a pony hired from a livery stable. I was assured that it was only sixty miles to the Fort, and that the pony could easily take me there in a day, but I soon found that he was just off grass, and weak and thin. I also discovered, after night had overtaken me, that I had been put on the wrong cattle trail. I reached a house in the evening, that of a school-teacher, who, because of his having had some education and possessing the ability to talk intelligently, was known in that region as "Windy" Turner, in distinction from "Bull" Turner, a cowman. I found him to be a gentleman.

The next morning he gave me directions as to how to reach the old trail that led to the Fort. I was to go to Wagoner's cattle camp, where the trail crossed Beaver Creek, and spend the night there. I traveled nearly all day, and reached the ranch building, the only house I had seen since I left the school-teacher's, only to find the camp deserted. Not a man nor a cow was in sight. As I had had no lunch, I was very hungry, and this being my first visit to this region, I did not know where to turn for food and shelter. At last, however, I saw a horseman coming toward me from the northeast, and rode to meet him. He was a cowboy. I inquired where Wagoner had gone, and learned that he had left a few days before for the Indian Territory. I was told, moreover, that the nearest place at which I could get a meal was back on Coffee Creek, which I had left in the morning. When I complained of being cold and hungry and of not liking to sleep in my saddle blanket on the ground without supper, the cowboy replied that he had not had a morsel to eat for three days and that he had slept for three nights in his saddle blanket. After that I said no more.

I was unwilling to return all the way back to the hospitable roof that had sheltered me the night before, and continued my journey, with no expectation of coming upon a human habitation until I reached Red River the next night. It is hard to express my delight, therefore, when, upon reaching the divide between Beaver Creek and Red River, I saw a lot of tents, some distance to the right of the trail. I hurried to the encampment, and found that it belonged to the locating engineer of the Denver and Fort Worth Railroad. When I told the young man from whom I had obtained this information that I wanted to see the engineer, he grinned (I was not a very pleasant-looking individual, covered as I was with the dust of travel), but he opened the door of the tent and said, " Here's a man who wants to see you."

As the occupant of the tent came forward, I presented to him my letter of introduction from the

Secretary of War; and I saw the grin disappear from the face of my guide as the engineer shook hands with me cordially, and remarking, "That is a good enough letter of introduction for me," placed himself at my service. When I told him that my pony and I were hungry, he instructed the man who had expected to see me refused the courtesies of the camp to get up a good supper for me and to care for my pony. Then, inviting me to make myself at home, he entertained me royally, and after I had made a hearty meal, opened a bale of new woolen blankets, and provided me with a most comfortable bed in his own tent. I hope if Major J. F. Menette sees this story, he will accept at this late day my thanks for his kindly treatment.

The next night I reached the crossing on Red River, where I found a house and stayed all night. The next day, about nightfall I crossed Cach Creek, and saw at my right, in a bend of the creek, an elevated "bench" on which a tepee was pitched. There were two Indians standing about, one a large, fleshy, good-natured man, the other thin, with large, prominent cheek bones, a typical Comanche. A large flock of children ran out to greet me. I must confess that I felt a little uneasy at being so entirely alone and at the mercy of these Indians, but I made the best of it, and as several turkeys were lying on the ground, I told the good-natured man that I wanted his squaw to cook me one for supper. This she proceeded to do, removing the breast and putting it on a wooden spit which she stuck in the ground before a large bed of coals and constantly turned until the meat was done. This, with a cup of coffee which she made me and the bread crumbs from my lunch, gave me quite a meal. I was too hungry to be fastidious.

The Indians were roasting camus, the bulb of the wild hyacinth, which grew plentifully in the creek bottom. They had dug a pit five feet deep and three in diameter and kindled a fire at the bottom, using at least a cord of wood to heat thoroughly the surrounding ground. The ashes were then scraped out, and the walls plastered with a mortar of mud, over which green grass was thickly strewn to prevent the bulbs from burning. The bulbs were then put in and covered with grass and mud, and a fire built on top of them. The next morning they were done, and were as much relished by these Indian children as popcorn or peanuts by the whites. I tasted some. They had a sweetish taste, a little like sweet potatoes, but they were so full of sand that my teeth were not strong enough to grind them up.

I put off going to bed until late, as I dreaded sleeping in the high grass where I had left my saddle. But at last the children, who had been amusing me, went off to bed, and I decided to go too. I spread half my saddle blanket under me, and with my saddle for a pillow was just dozing off when I heard a rustle in the dead grass, and the thin Indian, whom I disliked, stuck his head almost into my face. He had something in his hands which he wanted to swap with me for some of my property, and the more I argued, the more determined he was to trade. He wanted my pony, my Winchester, everything I had, and I was afraid that he would take them whether or no. At last, however, he left, crawling through the grass as he had come; but I was just dropping off to sleep, when I heard the snake-like rustle again. I was getting mad by that time, and when the Indian parted the tall grass and peered through the opening, he faced the muzzle of my gun, while I told him with much vehemence that if he did not go about his business and let me get to sleep, I would bore a hole through him. This had the desired effect, and but for the cold, which wakened me often, I slept in peace the rest of the night.

I was wakened in the morning by a shot, and a wild turkey fell from a tree near where I had been sleeping. They were so tame and abundant that they roosted in camp. The jolly Indian was anxious to earn another quarter, and as I had ordered turkey for supper, he had concluded that I wanted one for breakfast. I was not quite so hungry this morning, and detected the Indian smell which is left on everything they touch; but I made a brave attempt not to show my disgust to my host.

After breakfast, as I started out for the trail, a boy of fourteen walked down with me and stood talking, with his hands tangled in my pony's mane. I had given him some tobacco, and he was smoking a cigarette which he had made with a dry leaf. At our feet the path divided and encircled a little mound of earth covered with buffalo grass. When the boy had finished his smoke, he threw the still burning stump into this dead grass, which was damp with dew and sent up a dense column of smoke. This was all done so naturally that I thought nothing of it until I got up on the level prairie, where I could see for miles ahead. As far as the eye could reach, column after column of smoke was rising through the still morning air. It was thirty miles from the crossing at Cach Creek to Fort Sill, yet when I presented my letter to Major Guy Henry in the office at nine o'clock the next morning, the first question he asked was "Did you leave the crossing at Cach Creek about sunrise yesterday morning?" And when I answered that I had, he said that probably about ten or fifteen minutes after I left the creek, the Comanche chief had received notice by smoke signal that one man was coming over the trail toward the Fort.

In coming to Fort Sill, I had inadvertently come from one department into another, and the major had no power to send men out of his department without orders from General Sheridan, the commanding general of the Army. So I had to wait at Fort Sill until the matter could be arranged.

The southern cowboys, who hated the army blue and the darky soldiers who were stationed at the Fort, were doing all that they could to irritate the officers. While the latter were at dinner and the soldiers off duty, a squad of cowboys would ride into the post across the well-kept grass on the parade grounds up to the flagstaff, and fire at the Stars and Stripes. Another of their tricks was to shoot off the glass insulators from the government telegraph lines which connect the Fort with the headquarters at Leavenworth and with the Department of the Gulf. They had just accomplished this piece of mischief when I arrived at the Fort, and before the major could communicate with General Pope, Commander of the Department of the Missouri, in which Fort Sill was situated, he had to send out the signal sergeant to repair the line.

At last, however, all was arranged, and by general order, Corporal Bromfield, three privates, a six-mule team, and a wagon with a white teamster, and fifty days' rations, were detailed for my use. I started out with this escort, elated by the knowledge that I now had men and means of transportation upon which I could depend.

It is indeed a lovely drive from Fort Sill to Red River. We were rarely out of sight of the impressive Wichita Mountains, which rise from a sea of green plains like an islet in a lake. We reached the river on the second day, and had a mile of sand to pull through. At one time I thought that we would go down in the treacherous quicksands, but our magnificent team of dark-colored mules and the skill of the teamster carried us safely over. I have since seen, in the sands of this same river, holes ten feet deep which had been dug to rescue wagons loaded with valuable goods, that had sunk down to bedrock during high water.

When we reached the beds of the Big Wichita, we worked both Indian and Coffee creeks, a few miles apart. Here at last, after so much toil and so many hardships, I found myself in the very center of the fossil-bearing strata, and secured a number of fine specimens, among them the great salamander *Eryops*, the wonderful fin-backed lizard *Naosaurus,* that peculiar batrachian *Diplocaulus,* and other forms.

On arriving at the fossil beds, I showed Corporal Bromfield where I wanted him to pitch my wall tent, and went into the field with Mr. Wright, in search of fossils. When I returned at night, I found that the corporal had pitched my tent on a level and his own A-tent as close to it as he possibly could. "This will never do," I said to myself. "Discipline will go to the dogs, if I allow such close companionship." So I ordered him to take down his tent and pitch it a hundred yards away, and to follow this rule in future. The soldiers were very indignant, but they obeyed orders. As a general rule I found that I could handle them, although there were a few breaches of discipline.

I was so unfortunate on this expedition as to have my tent burned, with nearly all my personal property. When the men got to the flaming tent, the first thing they did was to cut the guy-ropes and let it blow over. They then, at my request, brought water and threw it on the burning sacks that held the fossils. This saved the fossils, but to do so we had to let everything else go.

On the twenty-fifth of April, we started with our load for Decatur, the nearest railroad point. We took the Henrietta road, and camped on the Little Wichita, where, in the sandy shales of the Upper Carboniferous or Permian, we found a locality rich in the fossil flora of that region. We secured a number of large fern fronds, etc.

Wild turkey were, as usual, abundant. Lee Irving, one of the escort, killed a hen and gobbler, and gave us a change from our customary diet of bacon.

On the fourth of May, after a long journey, we plowed through the valley named, and well named, the Big Sandy, and passing through groves of splendid live oaks, pecans, water elms, and locusts, reached Decatur, the terminus of the Fort Worth and Denver Railroad. Here I delivered to the agent my precious load of fossils, which had cost me so much expense, labor, and anxiety, and set out on the return trip to Fort Sill; where, on the twelfth of May, after a journey without incident, I turned over my command to Major Henry. The next time I heard of this splendid officer, he was a brigadier general in command of Porto Rico.

CHAPTER IX

EXPEDITIONS IN THE TEXAS PERMIAN FOR PROFESSOR COPE, 1895, 1897

IN the summer of 1895, sixteen years after my last expedition for Professor Cope, I was employed by him to make further explorations in the brakes of the Big Wichita. My assistant and cook was a farmer, Frank Galyean by name, who lived on Coffee Creek on the Vernon road, twenty-five miles north of Seymour. I camped a mile above his house on the west branch of the creek at Willow Springs, a favorite camping ground, as it was one of the few places in which water was always to be found. To the west rose Table Mountain, a hill several hundred feet high, and mountains of the same height extended in a southwesterly direction to Indian Creek, about four miles from camp.

I worked for several weeks on Indian Creek and Coffee Creek with very poor returns, but on the nineteenth of September, Mr. Galyean, who was of a sanguine temperament, announced that he had discovered the complete skeleton of a huge beast. So, filled with high ropes, I followed his lead along the rough face of the mountains, until at last, when we were completely exhausted by the ruggedness of the way, he pointed out a pile of the weathered and broken bones of a species so common that they were not worth picking up.

Dropping in a moment from my hill of expectancy into a slough of despond, I turned homeward, Mr. Galyean, who was as disappointed as I was, leading the way to a short cut through a gap in the mountains. As he got on the trail, which had been made by animals on their way to the spring, he stooped and picked up something, remarking, "Why, here's a bone!" I took it, and was astonished to find it a complete skull, covered with a hard siliceous matrix from a heavy bed of Indian red clay, which was completely covered with concretions. I had never carefully explored this horizon, as I had taken it for granted that it was barren. And I suppose that other collectors had imagined the same, for although it was within a mile of Willow Springs, where Boll and Cummins and other collectors had camped through a series of years, I was the first to discover this deposit of extinct animals.

We followed the trail over a slight rise into an amphitheater a couple of acres in extent, and then over a higher rise into another, a little larger, carved out of the mountain side and entirely denuded of soil. These two amphitheaters proved to be the richest fossil beds I ever discovered in the Permian of Texas. I quote the following entry from my notebook regarding this discovery: "After finding the perfect skull discovered by Galyean, we at once got into the richest ground I have ever seen in these beds. I got a perfect skull, and Galyean another. We have worked too low, it seems. This rich bone bed is on top of the beds I have been working, at the heads of the ravines that cut into the face of the mountains. The concretions in which the bones are preserved are in red clay, and are of greenish and other colors."

In my excitement over this rich find, I forgot my disgust with Galyean for leading me on a wild-goose chase, forgot how tired I was, forgot my dinner, forgot everything, and set to work at once collecting skulls and bones. I remember that I filled my collecting bag with seventy-five pounds of skulls, from less than an inch to over eight inches in length, and all new to me and to science. This load I started to carry down the steep trail to camp, a mile away. The good-natured Galyean, when he saw me tottering under the load, offered to relieve me of my burden, but I answered with such vehemence that no one should touch it, that I would break my back first, that it was more precious than its weight in gold, that he gave it up and fled down the mountains to camp, so that he might at least have a warm meal waiting for me when I arrived.

How can any man who has not had the experience himself, realize the glory of my triumphal march down that rugged trail? Not Nebuchadnezzar, when his chariot headed the army that was carrying away the treasures of the Lord's house from Jerusalem, with the king of Judah, blinded and bound in shackles of brass, in his train, could have known a prouder joy than I did now over this discovery of a new region, in the very heart of the old, which promised so rich a harvest of rare fossil remains. This is an instance of an experience which has been very common in my life—when I have been most completely hopeless and discouraged, I have made my greatest discoveries.

Of the remarkable batrachians and lizards which twelve million years ago peopled the estuaries and bayous of the Permian ocean shores, I found, during that three months' expedition, forty-five complete or nearly complete skulls, many of them with more or less perfect parts of the skeletons attached, and forty-seven fragmentary skulls, ranging in size from less than half an inch to two feet in length; the whole collection containing one hundred and eighty-three specimens of the extinct life of the

Texan Permian. The American Museum, which secured this splendid material, was unable to describe and publish it then, while the results of my famous expedition to these beds in 1901 for the Royal Museum of Munich were at once described by Dr. Broili. Consequently the American Museum lost much of the glory which attaches to the description of new material. However, the Permian collection in the American Museum is now being worked out with results of great importance to science.

Encouraged by my success on this expedition, I set out with high hopes on January twentieth of the following year to continue my work for Professor Cope in these beds. On reaching my headquarters at Seymour, I succeeded in hiring an old man with a a team and wagon, and on the twenty-fifth of January, I made my first camp on Bushy Creek, ten miles north of Seymour.

Three days later I found what I believed promised to be a fine specimen of the ladder-spined reptile, *Naosaurus*, called fin-backed by Cope. A number of perfect spines were exposed, presenting the possibility of securing a complete specimen. I worked very carefully over this skeleton, hoping to take it out whole and in good shape. It lay in red and white sandstone, which easily disintegrated on the surface into shale-like flakes. The spines and

FIG. 33.—FIN-BACKED LIZARD, *Naosaurus claviger*.
Restoration by Osborn and Knight. (From model in American Museum of Natural History.)

FIG. 33.—FIN-BACKED LIZARD, *Naosaurus claviger*.
Restoration by Osborn and Knight. (From model in American Museum of Natural History.)

transverse projections, which terminate in rounded knobs, were all broken *in situ,* and were also flexed and tilted with the strata, so that great care was necessary in following them. They were about three inches apart. I numbered the spines 1, 2, 3, etc., not with reference to their natural position, but to the order in which I came to them. A good many of the rounded ends of the lateral spines were missing, having been washed down the slope. I hoped to find them later.

As I studied these remarkable spines, many of them, near the center of the body, three feet high, with the lateral spines alternating or opposite, I instinctively called the creature the ladder-spined reptile; and I cannot see how Professor Cope could have imagined that these spines had any resemblance to the mast and yard-arms of a vessel, and that there was a thin membrane stretched between them which caught the breeze and acted as a sail. Later discoveries show it to be a land animal. Professor Osborn's magnificent restoration of the *Naosaurus* is shown. (Fig. 33.)

As I have said, it was a long and trying task to take up the skeleton, as it was in thousands of fragments. If I had dug them up as one would dig potatoes, no one would ever have had the patience to put them together again. So I took up each spine in sections, wrapping say fifty fragments together, and numbering them No. 1, spine 1, package 1, etc.; so that when the whole collection came to be put together, the sections could be mended separately first and then joined to one another.

The broken condition in which I found the skeleton prevented me from realizing then how complete and valuable it was; but as I look now at the fine photograph of the mounted specimen,—the only mounted specimen of the *Naosaurus* in the world (Fig. 32), I can see that this expedition was indeed a success, in spite of the discouragement which I went through at the time.

After the discovery of the *Naosaurus*, I was obliged to spend weeks of work without results, growing more and more disheartened because I myself was fully persuaded that the search was useless. Professor Cope was convinced that there was a fossil-bearing stratum between the Permian and Triassic, which would yield an entirely new fauna, and he had reasoned out that this ideal bed must be located northwest of the productive bed already known, in the very region, in fact, which I had gone over with such care for the Museum of Comparative Zoology of Harvard in 1882, and found barren. I, therefore, protested as strongly as I could against making the trip; but he insisted, and his more powerful will won the day. So I was forced to spend a month of extremely trying labor at the head of

Crooked Creek and in the other creek valleys, northwest of the productive beds.

Here were thousands of acres of denuded bluffs of red clay, cut into fantastic shapes, often resembling old fashioned straw bee-hives or crumbling towers and battlements. As far as the eye could reach, they spread out along the divide in ever-varying shapes. The beds disintegrated easily into red mud. There were no concretions, although the rock was full of concentric rings, from the sixteenth of an inch to an inch in diameter, consisting of a round white spot with a red rim. The narrow dikes which cross the thick deposits of clay are filled with fibrous gypsum. Underneath the clay lie strata of red and white sandstone and compact concretionary rock, all barren.

But the discouragement which attended my unsuccessful search was only one of the trials with which I had to contend that winter. In the first place, the weather was against me. It snowed or rained continually, so that the ground was never dry, and I took up ten or fifteen pounds of red mud on each foot as I walked. I came down with a severe attack of grippe, too; and to make matters worse, my teamster, who was also my cook, took a particular dislike to my stove, which had been manufactured under my own supervision and had always proved satisfactory with other men, and insisted upon doing all his cooking in a trench outside the tent, so that I lost the heat which I might have had but for his obstinacy.

Every morning I climbed out of bed with aching bones, and started on my long tramp. At first I would hardly be able to drag myself along, but gradually, as I warmed to the work, I would move faster, until usually I got so far away from camp that I should not have been able to return for dinner without taking more time than I could afford, and so went without that meal. After working as long as I could see, I would return to my uncomfortable camp, to go through the same performance on the following day. I had suffered from fever and ague in the fossil fields of Kansas, and had supposed that it would be impossible to suffer more, but I found the grippe even more relentless than the ague.

To add to my worries, the people at my post office had taken in a family with a malignant form of sore eyes, and although I supplied them with curatives, they would get careless. The peevish old man whom I had employed gave me a great deal of trouble too, at one time threatening to leave me alone in the brakes. In general, my experiences with hired men have taught me the advisability of owning my own outfit, whenever it is possible. A hired man knows how helpless one is in the fossil fields without transportation, and takes advantage of the

Phila Mch. 16 - 76

Dear Mr. Sternberg

I have your double letter of Mch. 9th & 11th. By this time you have the draft I sent you. I am glad that you have struck fossils at the mouth of E. Coffee Creek, & hope that you will have good success.

Your first letter is very blue but you must remember that bad weather is not your fault; neither is it your fault if you find nothing when following my directions. In fact you have no occasion to be blue as to yourself, for you fill an important place in the mechanism of the development of human knowledge. Very few men pursue a more useful life than yourself, and when the final account comes to be rendered, you will have no occasion to be ashamed of your record. I have personally the highest respect for your devotion to science. The serious worker in science holds a high position among men, no matter what the great herd may say about him. They simply do not know, & their opinion is not worth considering.

So pure science of the pure kind does not pay much money in this world but some time (after us), there will be more demand for our wares.

Very truly yours

E. D. Cope.

power which that helplessness gives him; or he looks at things from the hired man's point of view, and if he can better his wages by leaving his employer, thinks that he has a perfect right to do so, even if he has made a contract to remain.

After working for weeks in accordance with Cope's instructions, although it was as useless as carrying bricks from one side of a yard to the other and back again, I returned, worn and discouraged, to the beds which produced at least a few fossils. I determined, moreover, to give up the field at the end of my contract, and go home, and wrote a despondent letter to Cope, asking to be relieved when the contract expired, as I needed rest. It was then that I received the letter which I publish here in facsimile, a letter which I shall always cherish, not only because it shows the very best side of Cope's character, but because it makes me feel that he realized that my life work could not be measured by money. It gave me at the time the kind of encouragement which I needed more than any other, and on receipt of it, although I was just ready to give up from exhaustion and homesickness, I decided to remain another month in those barren fields. Cope promised that he would never again send me into a field against my own judgment; and by having my own way again, I was so fortunate as to add many new specimens to the collection.

For I was rewarded, as I have always in my life been rewarded, for my many days of fruitless toil, by the discovery of a long stretch of beds whose brilliant metallic color, the result of a large amount of iron accumulated by a dank and luxurious vegetation, testified that they had once formed the mud at the bottom of a bayou. This old swamp proved to have been the habitat of countless salamanders, and thanks to this discovery I accomplished more during the last month of my stay in Texas than during all the rest of the time put together, leaving out of account, of course, the fin-backed lizard.

I take pleasure in showing my readers a splendid skull (Fig. 34) after Broili, both the palatine and superior exposures of one peculiar species of these salamanders, to which Cope gave the name *Diplocaulus magnicornis.* The eyes are far down on the face, but with a broad expanse of sculptured bone behind, terminating in two long "horns," fourteen inches across from tip to tip, which are merely the greatly prolonged corners of the back of the skull. There are three rows of minute teeth in the roof of the mouth, and a couple of occipital condyles. The vertebræ have a double row of spines down each side of the median line, and the body is long and slender with weak limbs. The head was the largest part of the creature. This species was the most common of all those which I discovered in the Permian beds.

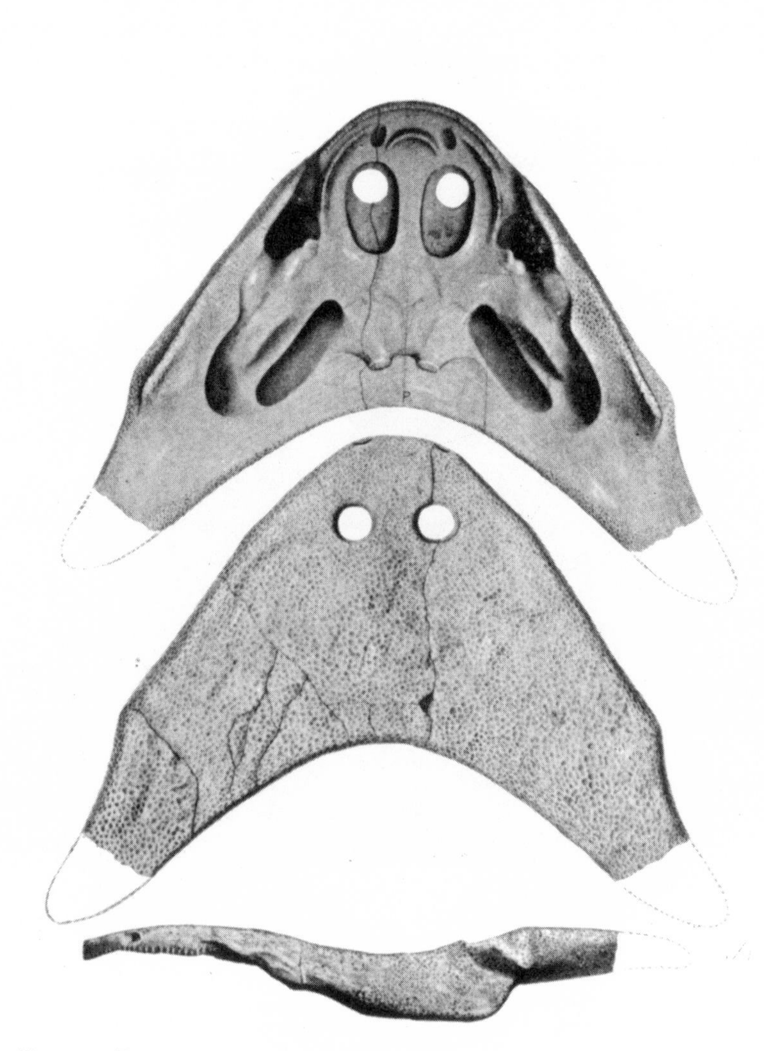

FIG. 34.—FOSSIL SKULL OF GIANT SALAMANDER, *Diplocaulus magnicornis.* Collected by Charles Sternberg in 1901. (After Broili.)

With best wishes for the successful prosecution
of your discoveries I am
Very sincerely yours
April 1902. Henry F. Osborn

FIG. 35.—PROFESSOR HENRY FAIRFIELD OSBORN.

Professor Cope used to call the specimens "mud heads," as they were almost always covered with a thin coating of silicified mud, which was very difficult to remove. In fact, nearly all the bones in this region were enclosed in a hard red matrix.

In the spring of 1897, I was again working in the Texas Permian for Professor Cope. He was deeply interested in the ancient fauna of the region, and I was sending him all the finer specimens by express, as I had during the last two years. On the fifteenth of April, I was camping on Indian Creek, having just completed a long and trying journey of about a hundred miles, around the Little Wichita and back to the main river at Indian Creek. During the trip we had encountered a terrible windstorm, which had threatened to carry away our tent, but we had weathered the gale and camped in the timber. I had gone to bed, but was roused from my cot by the arrival of a livery-man, who had been hunting for me all the day before. He handed me a message from Mrs. Cope, announcing the death of her husband on the twelfth of April.

I had lost friends before, and had known what it was to bury my own dead, even my firstborn son, but I had never sorrowed more deeply than I did now over the news that in the very prime of life, in the noonday of his glorious intellectual achievements, as he was bending all his energies to the study and description of the wonderful fauna of the Texas Permian, the greatest naturalist in America had passed away with his work undone. Death is terrible always, but it seems especially so when it strikes down men in the highest rank of intelligence, who are adding every day to the world's knowledge.

I was Cope's assistant in the field for eight seasons, and while we did not always agree, I consider the work which I did for him my most valuable service to science. It has often been my good fortune to supply him with some important link in the line of descent of vertebrate life,—such as, for instance, the famous batrachian genera *Dissorophus* and *Otocœlus,* reptiles with a carapace, indicating the line of descent of turtles from batrachians, or the camel from the John Day beds, with the metacarpals and metatarsals distinct,—and to furnish him with a large number of other forms which, with the material secured by his other collectors, helped him to acquire what Dr. Osborn has so truthfully called "a masterly knowledge of each type."

It is largely due to his efforts that the great science of paleontology, which, within my remembrance, had but few votaries, is now considered one of the most interesting studies of modern times. Well did he prophesy, "After us there will be more demand for our wares"; how well one can fully realize only when one remembers that the great

American Museum (whose department of paleontology under the able management of Dr. Henry F. Osborn (Fig. 35) is now one of the glories of science), that the Carnegie Museum in Pittsburg and the Field Columbian in Chicago and the Museums of Yale and Harvard and Princeton, besides many others both here and in Europe have been largely built up since he wrote those words. One thing is certain—as long as science lasts, and men love to study the animals of the present and of the past, Cope's name and work will be remembered and revered.

I am glad to be able to show a good photograph of this lamented naturalist (Fig. 15). Peace be to his ashes!

CHAPTER X

IN THE RED BEDS OF TEXAS FOR THE ROYAL MUSEUM OF MUNICH, 1901

ARNED by my experiences in the red beds of Texas without a team of my own, when I made a contract to conduct an expedition there under the direction of Dr. von Zittel of the Paleontological Museum of Munich, I resolved to ship my own horses and outfit to the field. I gave them into the charge of my son George, who was rapidly becoming a most valuable assistant, and saw him put them aboard a freight car and get in himself. The next time I saw him was at Rush Springs in the Indian Territory, on top of a freight car, skilled in all the lore of a brakeman.

We reached the old camp at Willow Springs on the thirtieth of June, 1901. The heat had already set in, promising the hottest season that I had ever experienced in the valley of the Big Wichita. It grew more and more intense as the months passed, the mercury often rising to 113 in the shade. All the water dried up in both the natural and the artificial tanks, and the short buffalo grass in the pastures curled up and blew away. We were camped in Wagoner's great pasture, twenty-five miles wide by fifty long, and I saw cattle die of thirst and starvation. Some had become so hungry that they had eaten the prickly pear, spines and all, and their mouths were full of putrefying sores where the spines had worked out.

The ground was hot, and the air like the breath of a furnace; and we had to haul all the water we used in camp from six to twenty miles. To add to our troubles, one of our horses, Baby, almost cut off her foot in a wire fence while striking at the flies, which, during the day, never ceased to torture man and beast. Even at night the horned cattle were not free from them, for they clustered around the base of the horns, fifteen or twenty deep, like hives of swarming bees, for rest.

The country was indeed a desert and deserted. All the people who had settled this valley on Coffee Creek or other streams, had gone never to return; the cowman had bought up all the homesteads. The schoolhouse in which I had so often attended worship had been moved from its foundations, and the houses that had once echoed to the merry cries of children, stood empty and desolate.

How can I describe the hot winds, carrying on their wings clouds of dust, which were so common that year and the next? I once went to Godwin

Creek, south of Seymour, passing on the way a hundred-acre field of corn. It belonged to an old man, who had cultivated it until it was perfectly clean, and the long rows of living green were beautiful to see. When I passed it again on my way back, a hot wind was blowing, so hot that I had to shield my face and eyes to keep them from burning. The beautiful field, upon which the old man had looked with so many hopes of a rich harvest, had been scorched and seared as if by a blast of fire.

So the weeks lengthened into months, and the merciless sky still refused us rain. At our camp on Coffee Creek the heat was so terrible that we could not keep eggs, butter, or milk, or many other edibles necessary to comfort and health. The result was that my stomach soon got out of order, and a severe attack of biliousness set in, attended by an incessant longing for a drink of cold, pure water. I thought by day and dreamed by night of the well on my farm at home, with the clear water dripping from the bucket; for our only drink, except coffee, was the warm, foul-tasting water which had been brought in a barrel from twenty miles away and had soon become stale. Even that was always giving out at inconvenient times. Whenever we came to a new fossil locality, and the hope was strong within me that now we would make a rich find, George was sure to say, "Papa, we're out of water," and we

PALAEONTOLOGISCHE
SAMMLUNG
DES STAATES.
Alte Akademie.

München, den 23 December 1901

Charles Sternberg Esqre.
Lawrence City Ka

My dear Sir,

Before receiving your last letter of the 6th Decemb. I had sent to your adress a cheque of 200$ as Salary for the last month of your collecting in Texas. I take notice of your freight expences (3$ 46c) and shall sent this little sum by an other occasion.

The 5 boxes with your great collection as well as the express box with the little skulls have been safely arrived. I have looked over the results of your researches and think, that the collection of this year is better than any other made before in Texas. With few exceptions we have nearly all the genera created by Prof Cope and several of them in much better condition. Beside theire is certainly a good number of very interesting and new material which will give us bussiness for several years.

I am very glad that I can give you such a satisfactory report about your hard work in the interest of our Museum and I hope to remain further in friendly relations with yourself.

With the best wishes for the coming year and the kindest regards

faithfully yours

Dr. Zittel

had to make the long journey through the awful heat over the dust-laden roads to the well at Seymour, twenty miles away. When we reached it at last, how we buried our faces in the bucket and the cool water!

But I will not dwell on this side of the picture, because there is another side. We were finding in wonderful abundance the material which we had come to secure, and the hardships were forgotten in the joy of success. In spite of the many obstacles with which we had to contend, we secured the collection described in that great letter from Dr. von Zittel which I publish here in facsimile and which I prize more than any letter I ever received.

Before I accepted von Zittel's offer that I should conduct an expedition for him in the brakes of the Big Wichita, I wrote to him, telling him how my work for science had had, from a material standpoint, no great returns. My life, I said, had been a constant struggle to secure sufficient funds to carry on the work, and the men who had bought my material had for the most part felt that they were doing good service to their museums by securing it at the lowest possible price, without taking into consideration that even a fossil hunter has to live.

It was with pleasure indeed that I received the answer of this great German, whose works on paleontology are used as text-books in our universities.

Dr. von Zittel wrote: "I am sorry that from your letter you do not consider yourself in a position to work for the Munich Museum in Texas this spring. I can readily understand that after your long activity in scientific fields without material results you are somewhat discouraged and embittered, and feel that your services in this direction have not been sufficiently appreciated. For my part, I have done my best to give you credit for the scientific side of your work, and your collections from Kansas and Texas in the Munich Museum will always be an everlasting memorial to the name of Charles Sternberg."

Such a letter, from a man like von Zittel, put new life and courage into my veins, as a similar letter from Professor Cope had once before, and made me feel that a little suffering more or less mattered nothing when measured with such enduring results. Cope is dead and von Zittel is dead, so far as such men can die, but I have preserved their letters as heirlooms for my children's children; for they testify that "no matter what the common herd may say about me," I have accomplished the object which I set before myself as a boy, and have done my humble part toward building up the great science of paleontology. I shall perish, but my fossils will last as long as the museums that have secured them.

But to return to the Texas Permian. I will follow my notebook for a while, as that, perhaps, is the best way to give my readers an idea of our life there.

On the eleventh of July I was in Seymour. I write: "A big dust storm struck the town, and this evening a rain is falling. This is indeed a great relief to me, as it will make the air cooler and give me water in the brakes, so that I can visit localities I could not before. My wagon, brought from Kansas, is a narrow-gauge one, and all the roads in Texas are cut by broad-gauge wagons. This forces my team to pull with one set of wheels in the rut and the other outside. Consequently the labor is wearing them out, in connection with the awful heat. I am, therefore, having new axles made, a long and tedious work, and I am resting out of the heat. Jesse S. Williamson has told me to occupy the building owned by himself and Will Minnich. It is a little cabin within a mile of the bone bed near Willow Springs. It has a tank of water for the horses, and is but a mile away from the schoolhouse, where a well has been dug. A few bucketsful a day, enough for camp use, trickles into it." This cabin proved to be a great accommodation, especially as the owners had a stack of sorghum, which was placed at my disposal and saved me the trouble of hauling out hay.

As one of my spindles was broken, I had to send

to Lawrence for another, and it was not until the sixteenth that I got my wagon from the shop. I then drove out to my old camp on Grey Creek in Mr. Craddock's pasture. Here, too, was the center of a field from which I had reaped a rich harvest for Professor Cope.

On the seventeenth, my notebook states that I was in the field all day and found fragments of skeletons and skulls, all broken to pieces and mixed up together. I could not find the horizon from which these specimens came. They were all piled together with concretions in a long, narrow wash, while above there was a level denuded tract covered with concretions. The only way in which I can account for the mixture of fragmentary specimens is that a bone bed lay above the level stretch, and in the disintegration of the deposit, the fragments were carried by floods into the narrow gulch, until not a sign of the original bed was left to mark its site.

I had sent a large collection from this same locality to Professor Cope, and he had been much interested, but had also been extremely tantalized by the fact that there were great numbers of fragmentary skulls, and that although the fragments looked freshly broken, none of the pieces could be united to form a perfect skull. I now found the same trouble again. Possibly some of the missing fragments of the skulls in Cope's collection, now in the American

Museum, may be in the lot sent to Munich, and vice versa.

On the nineteenth, I found the nearly perfect skull of a new species, and on the twentieth, another very fine skull near the locality from which I had secured the many fragments a day or two before. It was a skull of the great salamander, *Eryops megacephalus* Cope. There were six pairs of large teeth in the roof of the mouth, and a single row of various sizes in the mandibles. Some of the points had been broken off and were lost. The skull is over twenty inches long. All the bones are beautifully sculptured on the external surface. A few years before I had found a nearly complete skeleton of this creature, some twelve feet in length, lying at right angles to the Chisholm Trail. It was preserved in hard concretions, and had weathered out on the slope of a hill. The feet of countless cattle, just starting out on their weary journey for Kansas and the North, had worn away the solid siliceous envelope to the bones.

How the salamander tribe has degenerated since the days of these powerful creatures! Supplied with both gills and lungs, they dominated land and water, and increasing and multiplying in the tropical atmosphere, filled the swamps and bayous of this region. To-day we pull from some well or spring a weak creature called a mud puppy, and it is hard to realize that its ancestors, twelve million years ago, were strong and mighty, the monarchs of creation.

To return to Mr. Craddock's pasture; on July twentieth my notes read: "I am suffering from the heat, my tongue badly coated. However, I have got some splendid material. If I succumb to the awful heat and die, my discoveries will have done much toward enriching the collection at Munich."

On July twenty-first, I continue: "It is fearfully hot to-day, and I cannot work the beds without great suffering. I found a little skull."

The hot weather continued, and I went out to the cabin on Coffee Creek. Pet, our four-year-old, got away, and when George took her from a herd of horses, he found a big hole in her shoulder. "Both horses are failing fast," my notes read. "Have to send George in for feed. It is hard on the team to have to haul a load this weather through dust knee-deep, with no water fit to drink."

On the twenty-sixth, I was left alone, and went a mile north to the bone bed and began to dig into the face of a hard greenish layer of clay-stone, near a place where I had found some fragments in former years. I was delighted to find a pocket with two good skulls *in situ*, and the next day George returned with his load, and I had some fresh water, which soon, however, grew lukewarm. We found two more skulls in the pocket referred to, one of which was the *Labidosaurus hamatus* Cope, one of the earliest of reptiles. Another was that of a new genus and species which I found later, when we went back to Grey Creek to get a camp ready to receive Dr. Broili. He was to come directly from Munich to my camp in the red beds.

On the first of August, as we were out of provisions, we went into town. I rented a large room over a store building, and made tables and unpacked specimens for Dr. Broili's inspection. While I was working there, a storm of grasshoppers struck the building, beating against it like hailstones; and the next morning the ground was covered with them.

On the fifth, we drove out to our old camp on Grey Creek, and pitched two tents with the fly stretched between. The walls were elevated, and we were able to make a shade against the rays of the relentless sun. I went a couple of miles north, over the table mountain above camp, and found two extremely beautiful skulls of the long-horned amphibian,* *Diplocaulus magnicornis* Cope, a strange animal of which I have already spoken. I found also a specimen of the gar-pike, that ancient fish which has left its enameled scales in the rocks of many formations, whose descendants are still living in our rivers.

On the eighth of August, in spite of the debilitat-

*See Fig. 34.

ing heat, I started on a long trip to the head of Brushy Creek, on horseback. I climbed Table Mountain, which was, perhaps, three hundred feet above the camp, and struck west along the divide between the two creeks. I frequently left the horse tied to a fence, while I plunged down into the gorges on either side. At last, about three miles northwest of camp, at the bend of a branch of Brushy Creek, I noticed a denuded tract of the kind of bed I have already described, to which an abundance of bog iron lent a metallic luster; the very place to look for fossils.

The first thing I found was the perfect skull, six inches long, of a batrachian (*Diplocaulus copei* Broili); then, lying on the surface, another beautiful skull (*Varanosaurus acutirostris* Broili), with many of the bones of the skeleton, from which the hard red matrix had been washed off clean. The upper and lower jaws were locked together, and the long row of glistening teeth shone in the fierce light. The eyes were set far back, and the nose openings were near the front. It was so different from anything I had ever seen before that I was sure it must be new. Dr. Broili, in describing it, speaks of it as the most perfect specimen ever found in these beds. Nearly all the other skulls I had secured are compressed vertically, while this was compressed laterally.

I found in this bed hundreds of fragments of rock filled with the glittering scales of fishes, as brilliant now as in the days when they covered the bodies of these old fish. Here, also, I discovered a huge specimen of the long-horned species (*Diplocaulus magnicornis?*), and others much smaller, which proved to be the new *Diplocaulus copei.* "This," my notes say, "promises to be one of the finest localities I have found, and pays for the days of search under trying conditions."

When I reached camp, I found that George also had had a red-letter day, and had found a bone bed of minute animals on some brakes of Grey Creek under the roots of the grass in a washout. He brought in a skull, the smallest I had ever collected, with a great many broken bones and teeth. One specimen, which Dr. Broili named in my honor *Cardicephalus sternbergi,* was not over half an inch long. I secured here six skulls of the new *Diplocaulus copei,* also.

On Monday, the twelfth of August, Dr. Broili reached Seymour, and George and I met him at the station. A tall, strong, fine-looking German, with a full beard, he impressed me very favorably. The great difficulty was that, owing to my deaf ear, it was very hard for me to understand his broken English, and unfortunately I could not speak a word of German. I judged that he had learned his

English from an Englishman and not from an American, as he used a peculiar brogue with which I was not familiar. George learned to understand him better, and they became the best of friends.

We went back to camp, where we had the pleasure of Dr. Broili's company for two weeks, during which I formed a friendship which I have always deeply appreciated. He was delighted with my work and the material we had secured, but, as he says in the introduction to his great work describing my material, he could not stand the heat.

He describes part of my material in his splendid work on the Permian Stegocephala and reptiles, published in Stuttgart, with one hundred and twenty pages of text and thirteen fine plates. He says on p. 1: "The excellent results of the expedition of Mr. Sternberg in the spring of 1901 to Texas, which brought many very valuable specimens of *Eryops, Dimetroden,* and *Labidosaurus* to the Paleontological Museum's collection, caused the conservator of the Royal Paleontological Collection, Councillor von Zittel, to send out in the year of 1901 a second expedition to the Permian beds of the same territory, he being again successful in securing Mr. Charles Sternberg, the excellent collector from Lawrence, Kansas. Already in June of the same year he was in the midst of his sphere of activity in the Wichita Permian beds, near the small town of Seymour,

FIG. 36.—DR. KARL VON ZITTEL.
Born September 25, 1839. Died January 5, 1904.
(After Pampeckj.)

FIG. 37.—SHELL OF *Toxochelys bauri?*
Discovered by Charles Sternberg in Gove Co., Kansas. (After Weiland.)

Baylor County, located on a branch of the Fort Worth and Denver Railroad. On my arrival in the camp, through the assistance of the Royal Bavarian Academy of Science, it was made possible for me to take part in the collection from the beginning to the end of August. I found already a very good collection of very rich materials, which, besides parts of *Dimetredon, Labidosaurus, Pariotichus,* and other Theromorphs, included an excellent collection of different examples of *Diplocaulus,* of which some still possessed the greater part of the vertebræ. During my stay in that territory, our work principally consisted in making collections from our camp. We were compelled, on account of scarcity of water from the great heat, to keep near Seymour."

I am a patriot, and it would have pleased me to see all these splendid examples of ancient life enrich our home museums; but Germany is my fatherland, at least it was the fatherland of my fathers, and I am glad to have been able to build up there the best collection of Kansas and Texas forms in Europe.

One of the greatest prizes of the Munich Collection is a skeleton of *Labidosaurus,* now mounted there and collected by myself. *Labidosaurus* is important because it belongs to a very ancient and primitive group of reptiles, which, according to Prof. H. F. Osborn and other authorities, were the ancestors of all the later forms of reptiles.

After Dr. Broili left to return to Munich, I continued my work, camping on east Coffee Creek. Here again our search was rewarded. I found another bone bed of very small lizards, some of them, I think, not over six inches long. The skulls ranged in size from less than half an inch to an inch in length. Cope has given them the name *Lysorophus tricarinatus*. Drs. Broili and Case in their valuable papers have shown that this *Lysorophus* is one of the most interesting genera of all this wonderful fauna, since in the structure of the skull it is a veritable "missing link" between the batrachia and reptilia.

The deposit in which I found the *Lysorophus* was large, containing thousands of bones and many fine skulls. I am convinced that these creatures must have hibernated, as many of them were coiled in a circle in an envelope of hardened mud, and appear to have lain down never to wake again, each tiny reptile and its nest having been preserved through all the ages since. The flesh, of course, decayed soon after death, but by the process of petrification the bones have been replaced by stone.

Now I have always wanted to explain to a popular audience what this process of petrification really is. The word petrification should be dropped from our vocabulary, because it signifies an impossibility. I remember, as a boy, translating from the Latin a sentence like this—"His bones became stone," that is, turned to stone, and one often hears the expression petrified wood as meaning wood which has turned to stone; as if there were a process in nature by which one substance could be turned into another, as the philosopher's stone would have changed iron to gold. As a matter of fact, the process denoted by the word petrification is a process of replacement, not of transmutation. After the death of these ancient animals and the decay of their flesh, the water that passed through the bones carried from the cells of which they were made up the organic contents which decay, and left in their place deposits of the silica or lime which it held in solution. The same process continued when the lagoon bed was elevated above the water as solid rock. The rain-water, seeping down through rock and fossil alike, left in the bone cells the mineral matter it was carrying, until they were filled with it. Then, in process of time, the cell walls are broken down and rebuilt with silica or lime, and complete fossilization, or petrifaction as it is called, takes place, as in the case of the fossil bones in the Texas Permian. I found one specimen of the ladder-spined reptile in which the bones had been entirely replaced by iron ore, and others made up of silica.

How long does it take for the mineral matter to replace entirely the original bones? Ages upon ages. I found on the plains of Kansas a quarry of elephant bones, from which I took over two hundred teeth of the Columbian mammoth, some of the larger ones weighing fourteen pounds each. The broken bones were scattered by the ton through the matrix. I had them analyzed by Dr. Bailey, the head of the chemical department of Kansas State University, and he found only ten per cent. of silicified matter in them; that is, they were only ten per cent. less rich in phosphate of lime than Armour's ground bone meal. This great elephant lived about the time of the Ohio mastodon, whose bones have been found in such a position as to indicate that they were buried when Niagara Falls were six miles below their present site. So if we knew how long it has taken the river to dig six miles of its big ditch, we could tell how long it has taken to impregnate the bones of the mammoths in central Kansas with ten per cent. of silica. How foolish, then, to speak of completely petrified men, when man had probably not made his appearance in America at the time of the mammoths.

The rocks of the Texas Permian, as I have already mentioned, are of red clay filled with concretions of every conceivable form. I remember once rounding a butte and seeing before me hundreds of cocoanuts, some whole and others with the brownish shells broken, showing the white meat within. Absent-mindedly, I sprang from my horse to feast upon them, to find that they were concretions which had so closely imitated cocoanuts in shape and color that even I, an experienced collector, had been momentarily deceived. I knew, too, of a man who exhibited a collection of large concretions as fossil Hubbard squashes, and I heard no one doubting that they were all that their labels claimed.

There are two distinct formations in the Permian of this part of Texas which give character to the surface of the country. They are as different as if separated by hundreds of miles. I visited one locality on Pony Creek, where the red beds lay on top of the gray beds conformably. Looking to the west, a vast panorama, desolate and forlorn, of crumbling and denuded bluffs, narrow valleys, and beetling crags, spread out before me, with the usual red color dominant everywhere, its monotony relieved only here and there by the green of some stunted mesquite or patch of grass. To the east stretched the narrow valley of Pony Creek, whose topography is the same as that which is so familiar to the residents of eastern Kansas—a ledge of gray sandstone forming a narrow escarpment on either side and following the trend of the hills around the ravines, with grass coming down in gentle swells to meet it or rising to it from the bottom lands below. The greatest thickness of this sandstone, as I observed it, was at the head of a narrow gulch near my camp in the creek bottom, eight miles north of Seymour. I made a section there and sent samples of the rock to Munich.

I observed this rock under peculiar circumstances, and found that it solved an interesting problem—that of the water supply of the red beds. I discovered why the water that falls where these beds only are exposed runs off soon after a shower, except when caught in natural or artificial tanks, so that there are no wells or springs in the red beds, while in the gray beds there are always springs and streams of running water.

In the September of my 1901 expedition, the heaviest rain since May fell in torrents for an hour and a half; water lay everywhere on the surface of the ground. But soon after the rain stopped, it had all disappeared. My son had discovered across the creek a locality which was rich in fossil invertebrates, consisting chiefly of straight and coiled nautilus-like shells; and shortly after the downpour I went over to set to work collecting them, as Dr. Broili had told me that the Munich Museum was anxious to secure such a collection. I had not been long at work before George shouted to me that if I did not want to swim I would better cross the creek again at once. I followed his advice so hastily that I left my tools behind. Instantly, a raging, boiling flood of water covered the rocks in the bed of the

FIG. 38.—NIOBRARA GROUP, CRETACEOUS CHALK WITH CAP ROCK OF LOUP FORK TERTIARY, KNOWN AS CASTLE ROCK, GOVE CO., KANSAS. (Photo. by McClung.)

FIG. 39.—CHALK OF KANSAS, KNOWN AS THE COFFEE MILL. HELL CREEK.

FIG. 40.—BONES OF *Platecarpus coryphæus*.
As found by Charles Sternberg. Sent for mounting to Tübingen University.

creek, over which I had just crossed dry-shod, and rapidly rose to a height of eight feet, threatening to submerge my camp.

Looking for a good place to work on my side of the creek, the west, I found the gulch which I have referred to above. There was a level floor, formed by the first stratum of the gray beds, extending about five hundred yards to a ledge of red sandstone, eight feet thick. The floor was covered with débris washed from the red beds. To my astonishment, although the surface was dry, a flood of water was rushing out from under the upper deposits and tumbling in a miniature waterfall over the gray ledge, which was nearly five feet thick, into the ravine below.

The rock I found to be composed of four layers of sandstone. The upper layer, eight inches thick, is composed of fine-grained sand, which seems to have been ground to an impalpable powder by the beating of the waves. It is very compact and heavy, and upon exposure, breaks into rectangular blocks, so perfect in shape that they can be used for building purposes without being touched by hammer or chisel. The second layer breaks into large blocks of many tons' weight. It is coarser grained than No. 1, and is about twenty inches thick. It contains a few casts of invertebrate fossils. No. 3 is twelve inches thick, and is of the same general character as the other layers It is literally packed with casts of straight and coiled shells related to our living nautilus. They are mingled in great confusion. I believe some of the coiled shells are a foot in diameter. This stratum is not so compact as the others, and seems to contain more lime. No. 4 is a very solid gray sandstone, eight inches thick, its upper surface crossed at various angles by elevated ridges of harder material.

From these observations, I am led to the conclusion that the pervious nature of the red beds, which in the valley of the Wichita are about three hundred feet thick, allows the water to sink rapidly down through them until it reaches the impenetrable gray sandstone; from which it runs off at whatever angle the rocks may be tilted.

CHAPTER XI

CONCLUSION

I MAY begin this closing chapter by mentioning some other specimens which I have discovered, or which my sons have, for, thank God, I have raised up a race of fossil hunters. My second son, Charles M. Sternberg, has in his person recently fulfilled a dream of forty years of my own, by discovering the most complete skeleton known of Professor Marsh's great toothed-bird, *Hesperornis regalis*, the Royal Bird of the West. Unfortunately the skull is missing, otherwise the nearly complete skeleton is present, and strange to say in normal position, showing that Dr. F. A. Lucas is right in his restoration of the Martin specimen as mounted in the National Museum, i. e., as a loon, a diver instead of a wader, as had been supposed. Our specimen, however, shows a much longer neck than he had imagined. Strange indeed was this long-necked diver with its tarsus at right angles with the body and its powerful web-footed feet. The body was narrow, a little over four inches wide, with a backbone like the keel of a boat. The head was ten inches long and armed with sharp teeth. By keeping the body horizontal it could explore a column of water six feet high and wide, for any unfortunate fish within the zone of its activity. I would name this great loon the Snake-Bird of the Niobrara Group. This specimen I longed to find for so many years, but was glad to give the credit to my son. It is to be mounted in the American Museum, and I picture it as it left my laboratories (Fig. 41).

A word also about that great flying machine of the Cretaceous, the flying lizard *Pteranodon.* The skeleton and a very fine skull, which my son found on Hackberry Creek in 1906, is now mounted in the British Museum, where my warm friend Dr. A. Smith Woodward assures me "my specimens are greatly admired."

Especially have I been fortunate in the Kansas Chalk where my son, George Fryer, has charge as I write these lines of my twentieth expedition to those beds, and where he has discovered, and safely collected and shipped to my laboratory, a great plate of the beautiful stemless Crinoid *Uintacrinus socialis.* I sent one section to Professor M. Boule, of the National Natural History Museum of France, at Paris. Hundreds of these rare animals are represented in this slab (Fig. 42).

Before these pages go to press, and a year after I

Fig. 41.—Skeleton of *Hesperornis regalis*, the Giant Toothed-bird of the Kansas Cretaceous.

Discovered by Charles M. Sternberg. In American Museum of Natural History.

Fig. 42.—Slab of Fossil Crinoids, *Unitacrinus socialis*, containing 160 calyces, covering four by seven feet.

began work on them, I am pleased to be able to tell my readers of two noble specimens of the Pleistocene Age I have just secured from the plains of Kansas, that great treasure house of the animals of the past. One is a majestic Bison, whose head towering above that of his fellows supported a pair of horn cores measuring six feet from tip to tip. Along the curve the distance is eight feet. The length of the head is two feet, the distance between the horns sixteen inches, and from the center of the orbits, one foot. These splendid horn cores were uncovered through a fortunate chance. It seems that the Missouri Pacific Railway, wishing to shorten the creek in the vicinity of Hoxie, Sheridan County, Kansas, cut a new right-of-way for it across a bend. Their excavation came within two feet of the bones buried below, thirty-five feet from the surface of the earth; a friendly freshet washed them out, and they were discovered by Mr. Frank Lee and Harley Henderson, of Hoxie, Kansas, June 15, 1902. I was so fortunate as to secure them in June, 1908. I have filled them with white shellac, and they are now in condition to be preserved always, a specimen of the grand old bison of the Pleistocene time. Now their burial places are three thousand feet nearer the stars than the day they were buried there, as then the climate was semi-tropical and the land they roamed over near sea level. The largest pair of horn cores of a similar bison are preserved in the Cincinnati Natural History Museum. I copy from one of their records: "The most conspicuous figure on Plate IX, with immense horn cores, is of the long extinct broad-fronted bison. This specimen, by far the finest of its kind in existence, is the greatest prize in the Cincinnati Museum. It was found in 1869 on Brush Creek, Brown County, Ohio, and through the efforts of Dr. O. D. Norton it was acquired by the Museum in 1875." It gives me great pleasure to show my readers a photograph of the Kansas form that measures along the curve of the horn cores a foot and a half more than the famous Ohio specimen. (Fig. 43.)

The great Columbian Elephant, whose jaw I illustrate and have still in my possession, represents one of the largest, or the largest, of its kind ever discovered. It was found near the town of Ness City, in Ness County, Kansas. This giant lived at the same time the great Bison existed. The last molars have pushed out the worn premolars and the other two molars, and occupy the entire jaw, having a grinding surface of 5 x 9 inches. The lower parts of the teeth flare out like a fan, and measure twenty inches along the top of the roots. The greatest circumference of the jaws is 26½ inches, and the length 32 inches. Unfortunately, the articulations are worn away, likely by rolling in some river bed.

FIG. 43.—SKULL AND HORNS OF GIANT BISON FROM HOXIE, KANSAS.
Spread of horn cores six feet, one inch; length along curve, eight feet.

FIG. 44.—JAW OF COLUMBIAN MAMMOTH, *Elephas columbi.*
Discovered in Ness County, Kansas.

I secured this noble representative of American Elephants in June, 1908 (Fig. 44).

How rich are the strata that compose the earth's crust only a fossil hunter can fully realize. Take, for instance, western Kansas, where the soil beneath our feet is one vast cemetery. I know of a ravine in Logan County which cuts through four great formations. The lower levels, of reddish and blue chalk, are filled with the remains of swimming lizards, with the wonderful Pteranodonts, the most perfect flying machines ever known, with the toothed bird *Hesperornis*, the royal bird of the West, and the fish-bird *Icthyornis*, with fish-like biconcave vertebræ, with fishes small and great (one form over sixteen feet long), and huge sea-tortoises. Above are the black shales of the Fort Pierre Cretaceous, thousands of feet of which are exposed in the bad lands of the upper Missouri. In this formation the dinosaurs reign supreme. Still higher are the mortar beds of the Loup Fork Tertiary, where the dominant type changes from reptiles to mammals. Here, in western Kansas, are found great numbers of the short-limbed rhinoceros, the large land-turtle, *Testudo orthopygia,* several inferior tusked mastodons, the saber-toothed tiger, the three-toed horse, and a deer only about eighteen inches high. Higher still, where the grass roots shoot down to feed on the bones, are the Columbian mammoth, the one-toed horse, like our species of to-day, a camel like our South American llama, and a bison far larger than the present species.

The living bison has become almost extinct itself, through the agency of man. And in the layer of soil which covers all these formations, an old arrow-head and the crumbling bones of a modern buffalo give an object lesson in the manner in which these relics of the earlier world have been preserved. So races of animals, as of men, reach their highest state of development, retrograde, and give place to other races, which, living in the same regions, obey the same laws of progress.

My readers will be pleased, I am sure, to know that just before these pages go to press I am permitted to tell the story of our last great hunt in Converse County, Wyoming, during July, August, and September, 1908, for the largest skull of any known vertebrate, the great three-horned dinosaur, *Triceratops* (Fig. 45). Only thirteen good specimens are known to American museums, 7 of which are in Yale University Museum, and were collected, I believe, by J. B. Hatcher. From his field notes Mr. Hatcher has made a map of this region with crosses to indicate the localities in which skulls have been found, and 30 are so indicated, but I soon learned that he noted broken and poor material, as well as the more perfect. With my three sons I entered the region

FIG. 45.—THREE-HORNED DINOSAUR, *Triceratops* sp.
Restoration by Osborn and Knight. (From painting in American Museum of Natural History.)

FIG. 46.—DUCK-BILLED DINOSAUR, *Trachodon mirabilis*.
Restoration by Osborn and Knight. (From painting in American Museum of Natural History.)

with enthusiasm on the hunt for one of these skulls for the British Museum of Natural History.

I was not employed by that institution, but the agreement was, in case I secured a good specimen, it was to go to them. I must acknowledge I felt rather dubious when Dr. Osborn of the American Museum wrote me that he had had parties in these beds four years, searching without success for a specimen. For weeks and weeks we four examined every bit of exposed rock in vain. The rock consisted of clay and sandstone, the latter both massive and cross-bedded. Scattered through the great deposits of sandstone were peculiar-shaped masses of very hard flinty rock, with the same physical characteristics but with superior hardness. These added strange forms to the land sculptury. Almost every form the mind can imagine is found here, from colonies of giant mushrooms, to human faces so startling as to secure instant attention from the observer. (Figs. 38 and 39.)

A general view of the country from an elevated butte shows many cone-like mounds, resembling table mountains or even haystacks in the hazy distance! As the rocks, and even the flint-like material, readily disintegrate, the creeks that run east into the Cheyenne River soon radiate like the rays of a fan and deeply scar the narrow divides into rather deep canyons and narrow ravines. Perhaps a thousand feet of these fresh-water beds, are laid down in a basin surrounded, on all sides, by the marine, Fort Pierre, and Fox Hills Cretaceous.

Buck Creek on the south, Cheyenne River on north and east, and a line through the mouth of Lightning Creek would roughly give the area of the Laramie Beds we explored. They cover about a thousand square miles. Here in a country given up entirely to cattle and sheep ranges with but little of the country fenced, meeting no one but now and then a lonely sheep herder, my tribe of fossil hunters entered with bounding hope that we might find some of these famous dinosaurs.

Here is the border land between the Age of Reptiles and of Mammals, where mammals first appear as small marsupials. We secured several teeth of these early mammals. Day after day hoping against hope we struggled bravely on. Every night the boys gave answer to my anxious inquiry, What have you found? Nothing. Often we ran out of palatable food, as we were 65 miles from our base, and did not always realize how our appetites would be sharpened by our miles of tramping over the rough hills and ravines. One day in August, Levi and I started in our one-horse buggy to a camp we had made near the cedar hills on Schneider Creek. As we passed a small exposure which I had not gone over, I left him to drive and went over the beds of reddish shale, the remnant of an old peat-bog. I found the end of a horn core of *Triceratops*, and further excavation showed I had stumbled upon the burial place of one of these rare dinosaurs. How thankful we were that after so much useless labor we had at last secured the great object of our hunt. It will prove a beautiful skull when prepared and mounted under the direction of Dr. Smith Woodward, Keeper of Geology in the British Museum, where so many of my discoveries have gone.

Unfortunately the skull was somewhat broken up, and one horn core is missing. But one side of the face with the large horn core, the back of the head, and the great posterior crest, seems entire, as well as large pieces of the other side of the face, and a fine specimen will be made of it. The total length of the skull is 6 feet 6 inches. The horn core over the eye is 2 feet 4 inches high; while the circumference in the middle is 2 feet 8 inches, and it is 15 inches in diameter at the base.

This was a fully matured animal. As the bony ossicles of the head armature are co-ossified with the margin and remain as undulations more or less sharply defined, I am inclined to believe that they are ornaments. They might assist a little in defense but not offense.

In the mean time my oldest son, George, told me of a region he had explored a half-mile from our camp near the head of a ravine. Here we had found a natural cistern full of rain-water, protected from the sun and cattle by a couple of great concretion-like masses of rock that covered it. Over the divide where I had found the great skull, between Boggy and the breaks of Schneider near its mouth in Cheyenne River, George took Levi and myself. The evening before, I took the skull in to Lusk for shipment. George pointed out a locality in which he had found a bone-bed, where we later secured many teeth of reptiles and fishes, scales of ganoid fishes, bones of small dinosaurs and crocodiles and the beautifully sculptured shells of turtles, Trionx, etc. As there was still a tract of a few hundred yards to be explored the two boys started to go over it, while I went to the bone-bed. They soon joined me with the information that they had found some bones sticking out of a high escarpment of sandstone. George had found part of the specimen in one place and Levi another part soon afterwards. I requested George to carefully uncover the floor on which the bones lay.

While we were taking in our skull, George and Levi ran nearly out of provisions, and the last day of our absence lived on boiled potatoes. But in spite of this they had removed a mass of sandstone 12 feet wide, 15 feet deep, and 10 feet high.

Shall I ever experience such joy as when I stood

in the quarry for the first time, and beheld lying in state the most complete skeleton of an extinct animal I have ever seen, after forty years of experience as a collector! The crowning specimen of my life work!

A great duck-billed dinosaur, a relative of *Trachodon mirabilis,* lay on its back with front limbs stretched out as if imploring aid, while the hind limbs in a convulsive effort were drawn up and folded against the walls of the abdomen. The head lay under the right shoulder. One theory might be that he had fallen on his back into a morass, and either broken his neck or had been unable to withdraw his head from under his body, and had choked to death or drowned. If this was so the antiseptic character of the peat-bog had preserved the flesh until, through decay, the contents of the viscera had been replaced with sand. It lay there with expanded ribs as in life, wrapped in the impressions of the skin whose beautiful patterns of octagonal plates marked the fine sandstone above the bones. George had cut away the rock, leaving enough to give the impression that even the flesh was replaced by sandstone, giving an exact picture of him, as he breathed his last some five million of years ago.

A more probable explanation, judging from the shape of the skin outline which covers the abdomen and is sunken into the body cavity at least a foot, is that the great creature died in the water. The gases forming in the body floated the carcass, which was then carried by currents to the final burial place. When the gases escaped, the skin collapsed and occupied their place; the carcass sank head first and feet upward, the former dragging under the shoulder as the body came to rest on the mud of the bottom.

Quite different indeed is this grand example of extinct life from the one restored and of which an ideal picture is given in this book (Fig. 46). In the first place, in the specimen we discovered the ribs are expanded, the great chest cavity measuring 18 inches deep, 24 inches long, and 30 inches wide. I have no doubt but that with lungs expanded to their full capacity, he often swam across streams of water in the tropical jungle in which he lived and died. Further, the front limbs are not mere arms, that never touched the ground, but were used in locomotion, as there are toes with hoof-bones, not so large as those of the hind feet but with the same pattern, and a divergent thumb, that had a round bone for its ungual. Consequently the animal could use the front feet as clumsy hands to hold down the limb of a tree from which he was cropping the tender foliage, or banners of moss. There were three powerful hoofs on each hind foot.

I do not question, in the presence of this individual, which is complete excepting the hind feet, tail and left tibia and fibula, but that the reptile often stood erect, supporting his ponderous weight while feeding on the leaves of the forest. But when it walked it used its front limbs as well. A remarkable character are the countless rods of solid bone that lay along the backbone in the flesh, and appear like ossified tendons similar to those in the leg of a turkey. Hundreds of ossified rods appeared, row after row, shaped like Indian beads, as thick as a lead pencil in the center and beveled off to a small round point. It has occurred to me that these were for defense; that when a great *Tyrannosaurus rex* leaped on his back, his powerful claws found no lodgment in the flesh on account of these bony rods that could not be penetrated. Thus our dinosaur would shake off his enemy.

How wonderful are the works of an Almighty hand! The life that now is, how small a fraction of the life that has been! Miles of strata, mountain high, are but the stony sepulchers of the life of the past.

How rapidly has the field expanded which I entered as a pioneer some forty years ago! In 1867 I knew only five paleontologists—Agassiz, Lesquereux, Marsh, Cope, and Leidy, with but few followers; while to-day, Harvard, Princeton, the

American, the Carnegie, the Field, and the National Museums have all built up great collections of the animals and plants of the past, and the number of publications on fossil animals has reached an enormous total.

I had the pleasure of attending the meeting of the American Association for the Advancement of Science that met in the American Museum in New York at the mid-winter session in 1906. Professor Osborn introduced me to his splendid Head Preparator, Mr. Hermann, who has mounted the skeletons of the great *Brontosaurus, Allosaurus,* and so many other examples of extinct animals. Mr. Hermann was requested by the Professor to devote all his spare time to showing me anything the exhibition and storerooms contained, prepared or unprepared, and to do all in his power to make my visit pleasant. I certainly felt at home in that paradise of ancient animals, many of which I had collected for science on my own explorations. The magnificent halls in which they are exhibited are a wonderful tribute paid by the wealth and intelligence of the citizen of Greater New York to science. How admirable that Mr. Jesup should use his private fortune as the means to take from the obscurity of the private dwelling of the late Professor Cope his great collection, to which I was a contributor for eight years; and he has placed it under

Professor Henry F. Osborn, who with the assistance of Drs. J. L. Wortman, W. D. Matthew, and others, has brought order out of chaos and presented in intelligible shape not only that collection but many others from the fossil fields of the West.

It is a glorious thought to me that I have lived to see my wildest dreams come true, that I have seen stately halls rise to be graced with many of the animals of the past that lived in countless thousands, and that I have had the pleasure of securing some of the treasures, in the shape of complete skeletons, which now adorn those halls.

I stood on Columbia Heights that same year of 1906, and my heart swelled with pride when I looked down on that teeming metropolis and remembered that I too was a native of the Empire State. Then I thought of my distant prairie state of Kansas, and gloried in the thought that the best years of my life had been spent in her ancient ocean and lake beds, those old cemeteries of creation.

That past life, at least a very small fraction of it, I have sought to bring before my readers with pen pictures. We have men among us who can put their conceptions of the ancient inhabitants of land and sea and air on canvas, and among them are Mr. Charles R. Knight, of the American Museum, and Mr. Sidney Prentice, of the Carnegie Museum. Mr. Prentice I knew as a boy, and he has done me the honor to assure me that my words of counsel have done something at least toward assisting him to make the choice of following the work not only of an artist in a paleontological museum, but in portraying with pencil and brush the ideal pictures of the early denizens of earth as in life. His success is shown in his restorations of *Clidastes*. The results of Mr. Knight's restorations of many of the extinct animals brighten my pages, thanks to my friend Professor Henry F. Osborn, so if I have failed in my pen pictures to take my readers into the misty past, these brilliant restorations will certainly have the desired effect.

I cannot hope in this short space to have given more than a passing glance at the life of a fossil hunter. It has been one of joy to me; I should not like to have missed making the discoveries I have made, and I would willingly undergo the same hardships to accomplish the same results. And if my story does anything to interest people in fossils, I shall feel that I have not written in vain.

When I requested Professor William K. Gregory of Columbia University to be the final reader of the manuscript of this book, "The Life of a Fossil Hunter," shall I ever forget his kind words? "I hope you will not feel that you are under any personal obligations whatever, because this slight service is simply laid upon me by the necessities of the case, i. e., by the fact that your whole life and work have placed all paleontologists under lasting obligations to you." Surely "my cup runneth over; I have a goodly heritage." Greater than their obligations to me, are mine to the men of science who have described, published, but, above all, have prepared and exhibited the noble monuments of creative genius which I have been so fortunate as to discover and make known to the civilized world. My own body will crumble in dust, my soul return to God who gave it, but the works of His hands, those animals of other days, will give joy and pleasure "to generations yet unborn."

FINIS

INDEX

NOTE

In the preparation of this introduction, the author was fortunate to be able to consult portions of Katherine Rogers's unpublished biography of the Sternberg family. Several other sources were helpful, notably Richard Bartlett's *Great Surveys of the American West* (Oklahoma, 1962), Edwin Colbert's *Men and Dinosaurs* (Dutton, 1968), and Charles Schuchert and Clara LeVene's *O. C. Marsh, Pioneer in Paleontology* (Yale, 1940). Thanks also to David Dilcher and Elizabeth C. Raff for their helpful comments.

CHARLES H. STERNBERG, 1850–1943, hunted dinosaurs for most of his long life, often in the company of his three sons, George, Charlie, and Levi, dinosaur hunters in their own right. Charles continued his memoirs in a second book, *Hunting Dinosaurs in the Red Deer Valley*.